To

Pastor Solmes

– Joseph Chronley

Anonby Adventures in Missions

Anonby Adventures in Missions

Life Lessons and Ministry Meditations
from 55 years of ministry

by Joseph G. Anonby

XULON PRESS

Xulon Press
2301 Lucien Way #415
Maitland, FL 32751
407.339.4217
www.xulonpress.com

© 2018 by Joseph G. Anonby

All rights reserved solely by the author. The author guarantees all contents are original and do not infringe upon the legal rights of any other person or work. No part of this book may be reproduced in any form without the permission of the author. The views expressed in this book are not necessarily those of the publisher.

Scripture quotations taken from the King James Version (KJV) – *public domain*.

Printed in the United States of America.

ISBN-13: 9781545634387

Joseph and Ellen Anonby

Dedication

To the LORD

Whose mercies endure forever. Only by His grace and empowering has all that is eternal been accomplished.

To my WIFE

Ellen whose patience and godly counsel has kept me going through hard times and good times. She is the queen of our home.

To my PARENTS

Gilbert and Ruth Anonby came from Norway, looking for a new life and new opportunities. Dad always wanted to be a missionary to Africa. His eldest son, John, fulfilled that vision, and his youngest son, Joseph, became a missionary to Spanish nations.

Joseph G. Anonby

To my SIBLINGS

John August, my eldest brother by 4 ½ years, helped edit this book and support us in our missionary endeavours. For many years Dr. John Anonby taught English literature in Trinity Western University, Ft. Langley, B.C. Daniel Erling Anonby, (deceased), planned to be a doctor, but became a social worker and Anglican pastor in Delta, B.C. David, two years older than I, was a steady influence, supporter of missions and Gideon worker. All my brothers graduated as valedictorians from Central and Summit Bible School.

To my CHILDREN

Stan (wife Sandy), our eldest son, is working in Malaysia with Wycliffe, a Bible translation organization. Steve (wife Lori), has ministered as a youth pastor and is ministering to youth at risk. Joy (husband Ken), have been pastoring in Richmond for some years but are praying for Ken's health. All are Bible School graduates. They were all a great help to us in starting churches.

To Samuel Sieb

Samuel is a computer expert who helped us greatly in getting this book organized in a form that was satisfactory for the publishers.

Table of Contents

Chapter 1: Family History in the Making 1
Chapter 2: Launching Into Pastoral Ministry . . 26
Chapter 3: Preparing for the Mission Field . . 38
Chapter 4: Living in Argentina, our
Mission Field 49
Chapter 5: Starting a New Church in
Jujuy (1973-74) 62
Chapter 6: University Studies and Ministry
in Canada (1974-76) 77
Chapter 7: Founding Churches and a Bible
School in Argentina (1976-80) 82
Chapter 8: The Daily Life of a
Missionary Family 101
Chapter 9: Resettling in Canada (1980-84) . . 123
Chapter 10: Starting a Spanish Church in
Saskatoon, and Revival in
Argentina (1985-90). 140

Chapter 11: Missionaries to Barcelona,
Spain (1990-93) 152
Chapter 12: To the USA and the Canary Islands
with the Assemblies of God
(1994-2001) 168
Chapter 13: To the Dominican Republic
(2003-present) 195
Chapter 14: Beginning the "Deep River
Church" in Sto. Domingo 220

Chapter 1

Family History in the Making

A. Family Foundations from Generations Past

In the last five generations, the Anonby/Kasa family produced 18 missionaries and 25 pastors. My great aunt, Else Kasa listed these facts in a 45 page family history. It was written in mimeographed form, with multiple copies, about 1975. Since that time, several more have been added to that list.

What was the triggering event that began this chain of ministry? The Norwegian "Haugian (hill) revival" in the early 1800's was the starting point. The Norwegian version of Lutheranism welcomed the "pietistic movement" within the Lutheran church. Much of this evangelical and holiness move took place in home meetings. Years later, my father remarked "Nobody got saved in the formal Lutheran church. They got saved in the house meetings."

One of those saved and revived was Pastor Olson, who married Sidsel Kasa. At a family funeral in 1835, he preached a powerful sermon on the need for repentance before it was too late. It got results! Six of the seven Kasa children were saved! (I believe the seventh also came to faith later).

Soon several Kasas became ministers and lay preachers who rose to challenge their generation. Some Kasas went to Canada (Alberta and Saskatchewan), and to the USA (N. Dakota and Minnesota). One was a founding member of the Red River Seminary in Minnesota. Another preached and spoke out in a Lutheran religious magazine.

"I give people the law!" he proclaimed. "When men see their utter sinfulness, they have no excuse, no place to hide. Then I preach Jesus as the answer!" He also was devastating in his criticism of church officials who clung to traditions instead of confronting sin and preaching salvation. The church, in his view, had lost their determination to preach the whole gospel.

Years later, I studied for my Master of Divinity in the Lutheran Seminary in Saskatoon at the University of Saskatchewan, Canada. Professor Leesburg opined that "The Kasa preachers were sworn opponents of the formal Missouri Synod Lutheran church." He was not too impressed that I was a Kasa relative!

Now, in 2018, 183 years later, I am the fifth generation of that family who is still serving the Lord. Jillian Anonby, (our son Stan's daughter and my grandaughter), and Robyn Kutney (our daughter Joy's eldest girl) are both in ministry. Another grandaughter is in Bible School. This is now the seventh generation serving the Lord, and we praise the Lord for His favour.

But how did one scion of the "Kasa" family become the "Anonby" family? In Norway, I visited the farm where my father, Gilbert Anonby, grew up. His dad, Johannes Kasa, had farmed the "Anonby" farm for over 20 years. According to Norwegian law, he could now take the "Anonby" name as his own. This he did. Some time later there was a recession and Johannes lost much of his fortune. He moved to the town of Skarnes and lived there until his death.

"But grandma, you have no lap to sit on!" I was only six years old when in 1948 I visited Norway for several months with my mother. When we visited my paternal grandmother in Skarnes, I was overwhelmed by her sheer size. Close to 300 pounds! It made me afraid when she cajoled "Come sit on my lap!" During this Norway visit, I much improved on the Norwegian we spoke at home. At my mother's birthplace, a 250+ acre farm near Mysen, I met my maternal grandparents. They were God-fearing people who loved the land. Over 100 years later, the farm is still in the "Fjeld" family name. The Norwegian government has tax policies that encourage continuous family-owned farms.

I still have fond memories of sitting with Grandpa Fjeld on the horse-drawn hay rake. The farm had over 40 dairy cows and 6 working horses. I still remember most of those horses' names! I also remember attacking a wasp's nest in the barn. Six stings later, I ran to Grandma's house yelling "Wasps! They're still stinging me!" I felt "out of sorts" the next day. But it was hard to get me to bed. "It's still sunny!" I complained, since in Norway, the summer sun is still shining at 10 p.m.!

B. Enter My Father, Gilbert Anonby

Both Dad and Mom were from families with seven children. They were also the middle children, number four in the pecking order. Farming families sent their children to country schools, that mostly taught to the seventh grade. Dad recalls his school principal telling the visiting inspector, "That Gilbert is the smartest boy in school!"

The next step in Gilbert's education was to attend "Ungdom's Skool" (Youth School). There, youth from about age 15 were taught the Bible and proper social behaviour, similar to British "finishing school". Dad's eldest brother, Hans, had also attended, and there he met a fine girl called "Martha Fjeld". But more about that, later...

How do you make a living farming, when only the eldest male child (of seven) will inherit the family farm? Both my parents had siblings who decided to "go to try their fortunes in America (or Canada)." Dad's father, Johannes, spent some wonderful years in N. Dakota. "I only went back to Norway because I'd promised my Mother to do so" he told Dad. He had some dramatic experiences in N. Dakota, and told of a tornado that sucked up a small lake into a waterspout.

Dad was ready to travel and embark on a new adventure. In 1928, at the tender age of 20, he headed for Canada. For a while he worked for a stingy farmer near Swift Current, Saskatchewan. "I was so hungry, I used to eat wheat" he complained.

His brother Hans had been most reluctant to let his younger brother Gilbert, seven years his junior, to leave for Canada. Dad remembers him singing as they rode a horse cart "Og jeg skal liv og jeg skal du i Norge" ("I shall live and I shall die in Norway"). But, patriotism aside, it was hard to make a living in Norway. Grandpa was still running the farm, and Hans was his hired hand. So, he also emigrated to Canada, and joined Gilbert in Saskatchewan.

Soon they both decided to try their luck in British Columbia. There were a lot of gold and silver mines calling for workers. They bought an old car to drive across the Rocky Mountains. They must have heard some frightening stories. My uncle Hans carried a rifle, which lay beside their night time bedrolls. "If we are attacked by Indians, we'll shoot to kill" announced Uncle Hans.

Some Norwegian friends, the Johnsons, lived in Creston. For a while the two brothers lived there with them. Then they struck out on their own "stump farm" in nearby Arrow Creek. They worked hard at clear-cutting the 14 acres for farming. Times were hard, and food was often pancakes with syrup. Dad got so tired of syrup he swore off it for years. Later, when I was a child, he told Mom "Let's have pancakes and syrup every morning for breakfast!" "No sirree," Mom shot back "You'll be tired of it in no time! Remember Arrow Creek?"

After spending nearly seven years in Canada, some of it working in the mines near Nelson, Dad wanted to return to Norway to find an "Ekte Norsk frue" ("a genuine Norwegian bride"). Uncle Hans advised him to drop in on a girl, Martha Fjeld whom he'd met in the Youth School some years before Dad went. But first, he returned home and his father asked the classic question of the day, "Har du en gullur?" ("Do you have a gold watch?") Dad proudly showed his pocket watch, a symbol of prosperity.

Next on the list was the visit to Hans' old girlfriend, Martha Fjeld. She was a fine lady, but Dad's attention was drawn to Ruth, her younger sister. Ruth was 26 at the time, and had been jilted by the man she'd expected would be her husband. "I cried till there were no more tears to cry," she told me years later. She thought she'd be doomed to be an old maid, with her crooked front teeth. "But," she confessed, "He was not a Christian." I told God "If you bring me a Christian man as a husband, I vow to give an offering to the Lord in gratitude." (More about that later).

Gilbert showed up, and after he fell in love with a "genuine Norwegian girl", he held her hand and said firmly, "In one year we'll be married!" Dad was a man of his word, come what may. "Swear to his own hurt and change not" comes to mind. They had a great wedding, but Dad complained to me about the bedroom. "It was so hot with that little room stove, that we had to sleep with the door wide open!" He was still indignant. Years later I slept in those same upstairs bedrooms, each with a tiny stove, in a time when few homes had central heating. And winters are cold in Norway!

C. The Anonby Family begins in Canada

Honeymoon on a boat to Canada! That sounded exciting, and it was often a 10 day trip. But for Ruth Anonby, there were many tears. She was leaving her family, her friends and her native country. Life as a miner's wife in British Columbia was tough. The gold-mining town of Sheep Creek, (now a ghost town near Nelson), was tough. Four children were born in 4 1/2 years. Apple boxes served as chairs in a little shack. That shack was later hauled to nearby Nelson, and became someone's garage.

They had some harrowing experiences. "Bear!" cried the little tykes, and wrapped themselves around Dad's legs as he was picking huckleberries. John was run over by visiting Uncle Hans, but the Model A center went over him, not the wheels. Daniel fell off his apple box seat and broke his hip. (The first of a long series of broken bones over the years!)

Long term, the most serious problem was a lung disease called "silicosis", from the silica dust in the mine. Mining companies fought against any liability for anything short of tuberculosis. Dad had silicosis, but experienced a dramatic healing. Years later, both he and uncle Hans were interned for many months with tuberculosis. Finally a (little) compensation was paid out, based on salaries paid many years previously.

"More, more about Jesus, more, more about Jesus..." That became Dad's favourite song in the house meetings he attended. Mrs. Jones, a former Salvation Army officer, had become a firm Pentecostal believer. She loved Jesus fervently, and would go on long fasts. She said "Jesus would come to me in the night and spoon-feed me nourishment." I don't understand it, but that is what she reported.

Dad received the baptism in the Holy Spirit, and spoke in tongues. At times he would sing in the Spirit. I remember hearing his fervent prayers coming up from our dirt floor basement where he'd kneel by an old car seat. Mom was a traditional Lutheran, and had been warned against those Pentecostals. "Stop that devilish nonsense!" my great-grandfather, a Lutheran preacher, ordered when someone called out "Praise the Lord" during his sermon. For some time, Mom favoured the "Mission Covenant" church, of Scandinavian roots, in Nelson.

Shortly after I was born, in 1942, the family moved to Nelson. Dad bought a one bedroom house on four lots. I recall sleeping in a large steel crib in my parents' bedroom. But I was soon moved to the very low-roofed attic. We climbed a steep ladder to get to bed. At times the bedbugs visited us and the attic would be fumigated.

But before bed, we had family devotions. We spoke Norwegian at home, with my brothers John, Daniel and David. One devotional book we read nightly was by Spurgeon (translated into Norwegian). Then Dad would read the English Bible. Choruses were usually in English and prayers, one child after another, were in Norwegian. When I was about six years old, all four of us boys got our own Bibles. Mom would awaken us every morning, and place the Bible beside us. (Sometimes she'd bop me on the head with the Bible to wake me up!)

By this time, Dad had added another bedroom to the house. Plus a goat shed which was cleaned out and turned into an additional "outside" bedroom. It was VERY cold in the winter, and sometimes even the "thunder-pot" would freeze. Dad talked for years about installing an indoor toilet, but never did. We lived on the town outskirts, and were not connected to the town sewage system. A wood stove helped preheat the 'goat shed', but with no insulation, the bedroom cooled off quickly. We'd sleep "double" to keep warm, and Mom had huge home-made quilts we'd pile on. I called them "weight without warmth."

D. Joseph grows, spiritually and physically

Church was a large part of my life. We walked about one mile to Sunday School, church, and later to Sunday evening and youth services. My Dad was Sunday School secretary/treasurer for many years. Rev. Ian Presley was the pastor I remember from my early years. There was no microphone system, but he had a LOUD voice! I remember trying to sing as loud as he did!

Woodrow Fletcher was a dynamic pastor, and the church grew from about 80 to over 150. Our Sunday School won national prizes for growth. And we enjoyed special meetings with a series of evangelists. Watson Argue, Junior was one of the men God used in a special way. The grandson of an early Pentecostal Assemblies of Canada pioneer, he preached with Pentecostal power. After two weeks of special meetings, 17 received the baptism in the Holy Spirit. That included our entire family, with myself receiving the last night.

My brother David prayed with me and encouraged me. I began to speak in tongues, a few words, and then would pause and speak again. In my enthusiasm I'd vigorously wave my hands at the wrists. Evangelist Argue gave me guidance twice that night. First, as he noted how long I'd been seeking, he suggested "Ask God to forgive you for any sins you've done." I couldn't think of any at the moment, but I swallowed my pride and prayed for forgiveness. Soon after I began speaking in tongues. Next, he suggested "Try speaking in tongues without waving your hands!" I tried, and continued to speak in tongues. My Dad had often challenged us boys "You're not Pentecostal yet", but now we were, ptl.

Ministry Meditation:

Many times over the succeeding years we've encouraged people to receive their "personal prayer language". When people feel God's power, their body often "manifests" a reaction to God's presence. They may cry uncontrollably, shake, fall or copy what they have seen others do. We find this is unhelpful and try to discourage it. Instead, when we discern that they are close to receiving, we instruct "now, just speak out in the language God is giving you". (More about this later)..

E. Hospitality "of angels unawares"

Our home was small, but it was a rare Sunday we didn't have someone for the big noon meal. Dad would invite new folk, especially single men. Over time, we had nearly every regular attender of our church over for dinner. Some would reciprocate. By this time we had a 1948 Chevy, and one of our frequent visitors, George Elsey, served as long term "border-guest" and chauffeur. The Hood family had three children, and Ken, the eldest son, was my age. We became fast friends, and were frequent Sunday guests at each others' house. We attended Christian Camp together, went to the altar to pray together, and vied for some of the same girls.

But my favourite guests were missionaries. Dad had told us "I wanted to come to Canada, make money, and go to Africa as a missionary." I think Dad wanted all of us to be missionaries. I loved to listen to missionary stories; they charged my imagination. In grade nine we were asked to prepare a book report of what we wanted to choose as a profession. I put a picture of Jesus on my report cover, and head-lined the page with "Go into all the world and preach the gospel!"

Life Lesson:

I believe an important reason for my becoming a missionary was my parents' hospitality. Children quickly adopt their parents' heroes. Also, if your home is a "safe and welcoming place" for youth, they won't need to find friends elsewhere. "Use hospitality without grudging" 1 Peter 4:9 is a biblical command. It is not just for "those who have the gift". We won more souls to Christ in our first church by hospitality than by any other method. Hospitality has been a major part of our ministry all our lives. We build relationships with leaders and we've begun churches in our home. Often we had little to share, but the main attraction is Christian love.

F. Off to Bible School

The family pattern had been set. Finish high school, take grade 13 (first year university) then to Bible School and on to university. By the time I was ready to graduate from high school, John and Daniel had already graduated as valedictorians in our Saskatoon and Vancouver Bible Schools.

"But," I reasoned, "Why learn all that complex math and science, then forget it while attending Bible school? I think I'll take one year of Bible School, then go to university." Besides, I didn't have too much money. My "summer job" was picking huckleberries and selling them to stores or friends. Our family had become quite proficient at that, picking thousands of pounds every summer. I averaged over 43 pounds a day, which I carried down the steep Hall Creek hills on my backpack.

Once I was by myself, age 12, and a mother black bear with two cubs (one brown, one black) began picking at the same bush. The mother bear sniffed the air, reared up on her hind legs, and gave a "woof". Her cubs scampered up a nearby high burned-out tree. My pack-sack with many pounds of berries was at the base of that tree! She reared up the second time. I prayed out loud, "leave in the name of Jesus!" The bear didn't hear my prayer. So I backed away slowly, turned and walked slowly, then...RAN LIKE CRAZY. I found my brothers and, panting, told my exciting story. For awhile they let me pick and put berries into their packs, but finally demanded "Go and get your own pack!" We were quite a competitive lot, with David, at 65 pounds picked in one day, the champion. But at the 10 cents per pound that Dad paid us, my savings were small.

We all loved the outdoors. Hiking, fishing and mountain-climbing were our weekend sports. When I was 10, John bought a "Kootenay Sports Map" depicting all the area's creeks and lakes. Our first mountain lake "full of fish" was Clearwater Lake, near Nelson. About a four hour hike up the creek brought us to a pretty lake. While I was cutting down my willow branch for a fishing rod, Daniel had already caught two fish on his "wiggly worm" bait. There were 100 creeks and nearly as many mountain lakes to explore. That became the mission of the "Anonby goats" as people began to call us.

It was an adventurous life, and at age 16 I shot my first buck. Yet Bible School was calling me. Mom was very reluctant to see her last child leave home. Dad lent me $100 to make up my expense shortfall. And, with Marilyn Stenberg, a girl from our youth group, I was "off to Bible School!" in N. Vancouver, B.C. The first night there we had a "get acquainted evening". Our freshman class numbered an impressive 42, the largest number in many years. I remember seeing a cute, vivacious blonde girl from Penticton. Her name was Ellen Bonderoff...and she sat in the seat in front of me.

Ellen had a unique story of her own...

A. Pacifists in Russia

"Burn your guns!" was the battle cry. Many of the common people had grown tired of the constant Czarist wars. Compulsory military inscription hung like a Damocles Sword, by a thread over the heads of simple farmers. They were tired of losing their sons and tired of fighting. "Toil and the peaceful life" became their motto. Encouraged by Tolstoy, the writer of **War and Peace**, and also by the Quakers, they petitioned to emigrate from Georgia, in Southern Russia, to Canada's wide plains.

A large group called themselves the "Doukhobors" – "wrestlers in the Spirit", instead of the flesh. And they DID burn guns! The problem with that was these were guns in the Russian government armoury, not ones in their personal possessions. A few decided to part with their possessions, including their clothes. The first of many "nude protest parades" began in Russia.

Canada was competing with the USA, Argentina and Brazil in welcoming new settlers to develop the vast lands of the west. The Doukhobors got assurances of freedom from military service for 70 years. During the years from approximately 1890-1920, between 20,000 to 40,000 sought a life with freedom from oppression in Canada. Among those they counted as oppressors was the Orthodox Church. As the state church for hundreds of years, it collected tithes, owned large tracts of land, and was a feared political power. Their priests were contemptuously called "Popes" by the dissenters.

The Doukhobors' Russian leader, Peter Veregin, was allowed to leave exile in Siberia and accompany them. Incipient collectivist/communist ideas at the time rallied around the ideal of equality and "collectively-owned property". As colonists arrived, they settled first in the town of Kamsack, east of Saskatoon, Saskatchewan. But the government, to the dismay of some leaders, only granted land titles to individuals, not an entire group. In response, a substantial faction broke away and petitioned for settlement land in British Columbia. They were following their motto of "toil and peaceful life". Much toil!

B. The Doukhobors in British Columbia

The Kootenay valley as well as the Grand Forks area was soon dotted with new settlements of farms. A large three storey brick house (called a "seeloh"), with ample bedrooms for several families, was commonly the centerpiece structure. Each family had a small plot that was "their garden", and they would work together on their common lands. But

Ellen, who briefly lived in one of these large houses, says "It didn't work. People would steal the produce of others living in the same house, or they'd destroy someone else's more productive garden."

There were more insidious and larger issues at stake in what appeared to be a peace-loving group. In Nina Holt's book **Terror in the Name of God**, she explores the internal leadership battles. A breakaway group, called the "Sons of Freedom" began a reign of terror. Nude marches were a common form of protest. But the anti- government movement grew ugly as dozens of schools were burned and bridges dynamited. Schools were said to "teach people to love war." Peter Veregin was killed aboard a train, with a satchel-bomb probably planted by a rival leader. But blame was deflected towards the government, so this justified blowing up train bridges. The rival group, "Sons of Freedom" also twice burned down a successful jam and fruit bottling factory at Brilliant, near Castlegar. This was a sabotage against the more regular Doukhobors, who had to defend their own homes from being burned down.

Peter Veregin was succeeded by his son, also called Peter Veregin. But he was an inveterate gambler, alcoholic and womanizer. He tried to get as many lands registered into his own name as possible. This was under the guise of "collective ownership", but in fact he then mortgaged the lands to pay personal gambling debts. He discouraged the simple preaching of the Bible, though Ellen recalls talk of "Ye must be born again." Literacy was not encouraged, because ignorant people are easier to control. I heard a distant uncle of Ellen give the party line years later "We seek the spoken word, not the written word." Over the years Doukhobor spirituality consisted mostly of singing songs and reciting psalms at funerals. I used to enjoy their multi-harmonied a capella choirs in our Nelson community Christmas festivals. A lead singer would pitch the song, and others would join in. They did not approve of organs and other such instruments. Those were disdained as being "too English."

C. Ellen comes to faith.

Ellen was the eldest of five children, born in Pass Creek, across the Columbia river from Castlegar, and 28 miles west of Nelson, down the Kootenay river. She was sent, somewhat reluctantly, at the age of seven, to a tiny one room schoolhouse. Her home language was Russian, similar to all of her friends. She learned English in class, but flunked grade one in the process. It didn't help that the lady teacher was an alcoholic, who was eventually fired. (Ellen "caught up" by skipping grade three, with a new teacher.)

Summer time was free time. And two young ladies, Misses Musikal and Lashenko came to teach a Daily Vacation Bible School. "They were so kind" was Ellen's recollection at about age six. The first seeds of the gospel were planted in her young, fertile heart. Life was hard for these pioneers. Her Mom had to break the ice to get river water to drink during the cold winters. Ellen recalls weeding what seemed like endless rows of vegetables and strawberries. She once succumbed to a sunstroke, and has been allergic ever since to excessive direct sunshine.

Her father, Nick, was a hard-drinking, hard-working sawmill owner at one time. The sawmill would whine loudly for hours close by the house. This damaged her hearing. They'd play in the huge sawdust pile. She recalls her Dad counting stacks of money. But tragedy struck one day, when her uncle Bill died instantly from a broken sawblade piercing his heart. Her Dad buried his grief in more drink, and soon lost his business. He was of Russian extraction, but never a true Doukhabor in terms of personal faith. Doukhabors were vegetarians, and "thou shalt not kill" also meant "Eat no meat". How can you eat something DEAD?" Ellen didn't eat meat till she left home. But she defied her Mom when she told her "Marry a Doukhabor." "Show me a Doukhabor who isn't a drunk!" she shot back. Their traditional faith had failed after a string of unscrupulous leaders, and drink became their new god for that generation.

The family moved to Grand Forks, where her father worked in a sawmill. And there Ellen found a little Brethren Church that faithfully sent someone to bring them to Sunday School. At times they'd have to wait as Ellen and her siblings got hurriedly dressed.

Soon it was moving time again, one step ahead of the landlord and away from more debts. This time the family, now with five children, moved to Penticton, on Okanagan Lake. Would you believe it? She moved next door to the Sunday School Superintendent of Bethel Church. This was the "Age of Sunday School", and they, like Nelson, had won awards for bringing in the children to hear God's Word. Contests and friendly competition built enthusiasm. A carefully organized followup meant that even the pastor was an integral part of this evangelistic arm of the church.

As Ellen grew a little older, she was invited to join the youth group. She really liked the parties, but also liked to dance and go to shows. She'd go to church with her friends, often sitting in the balcony with them as they chewed gum and passed notes. She loved the singing, and enjoyed their monthly parties, often on the nearby beaches. She also loved to hear missionaries speak. But she hated the "altar calls" because her heart beat so loudly. She was afraid a nearby friend would hear the thumps. It was too big a step to walk to the front. But one night, about age 14, after "counting the cost," she said "Yes" to Jesus as she walked home from church.

"I always had felt there was a limit, like a ceiling to my joy. But when I accepted Jesus that night, it felt like there was no ceiling, no limit to my joy!" That has been Ellen's experience ever since, and the Psalmist did say "In thy presence is FULLNESS OF JOY and at thy right hand are pleasures forevermore" (Psalm 16:11).

Ministry Meditation

Children are among the easiest to reach for Christ, followed by youth. It is remarkable how much children can remember, especially during the "golden age of memory," ages 6-9. Churches need to take full advantage of that. I was saved at the age of six in family devotions. Ellen was saved because of Sunday School and a youth group that had "fun activities" which drew her in. Good and wholesome events are magnets that eventually draw people to hear about Jesus. Then you win a "servant for life" for Jesus, without the scars and regrets that taint the (far fewer) adult conversions. Over 90% of missionaries were saved as young children.

Pastor John Peel, the brother of South African missionary Bob Peel, was Ellen's pastor. His sweet wife Sylvia was her Sunday School teacher. "Look how much easier it is to pull a person down than up" she'd say, as she had one student stand on a chair. Then she'd drive home the main point "The only way to keep standing for Jesus is to have personal devotions!" It was no easy task for Ellen in a small, rented house to find a private place. She'd often lock herself into the bathroom for devotions!

Ellen became a Sunday School teacher, and the social convener for the youth group. "We've lost you from the family!" her mother would wail, "You just go to church and love those people, you have no fun anymore!" "Oh Mom," she'd reply "If you only knew how happy I am now!" Her Dad threatened, "If you keep going to that church, you have to leave home!" Ellen considered sleeping on a park bench, but never leaving Jesus.

There were many sad moments living in a house with an alcoholic father. Alcoholics often "step on people" to compensate for feeling unworthy. Ellen grew up with an inferiority complex, but God healed her spirit after salvation.

One infamous Christmas her mother was crying, "There's no money for presents, no money for celebrating!" But there was a large box filled with expensive liquor in Daddy's bedroom. One night Ellen broke every bottle in that box. She had to hide from her furious father for several days. Money was always tight so Ellen went to work in the local cannery at nights to help pay the rent. She'd be so tired she often fell asleep in the tub. But Sunday morning she'd be up early for Sunday School.

Bible School singing groups would come to Penticton every summer. Soon Ellen had a new goal: Bible School. Her parents were horrified, and certainly not about to pay for the studies. She found a job in Woodwards, a large Vancouver department store, for a year. Now she was ready for Bible School and a new future! There was this smart guy called Joseph Anonby that sat behind her. But she refused his invitation to go out, and suggested he try someone else. She thought there were too many differences between us. Joseph appeared to be "too stuffy and rigid". Interestingly enough, a future sister-in-law, Rose Anne Giesbrecht told her "Don't think the Anonbys are so high and mighty. They still have an outhouse!" Another student, Faye Farnell, told her "Whoever marries Joseph is going to go places." At that time, Ellen had no idea just how many places that would be!

But before the third year of studies, Joseph stopped by at Cache Creek where she worked every summer at "Hungry Herbies." He kept stopping by, and by graduation time was looking for a ring. What had started as a good friendship had become a life relationship now well over 50 years!

**Anonby family outside church.
Joseph is the babe in arms.**

Joseph G. Anonby

Wedding, October 1963

Anonby Adventures in Missions

*Please Pray for Us
Your Ambassadors for
Christ in Spain*

**Rev. Joseph & Ellen
Anonby**
Foreign Address:
Seminario Evangélico
Apartado 31, 19200
Azuqueca de Henares
(Guadalajara), SPAIN

Home Address:
Box 515
Camp Hill, PA 17011
Phone (717) 761-4640
Fax (717) 761-0369

---

*Only prayer breaks
spiritual chains*

*"The weapons of our warfare
are not carnal, but mighty
through God to the pulling
down of strong holds."
2 Cor. 10:4*

PRAY FOR US
Missionaries to Argentina
THE ANONBYS

JOSEPH & ELLEN
Stanley, Steve, Joy

Come along
And build with me
Building for
Eternity
Winning Souls
We set men free
This is our work
Come work with me

Address:
Rev. & Mrs. J. Anonby
c|o 10 Overlea Blvd.
Toronto, Ont.
M4H 1A5

Early prayer cards

Chapter 2

Launching Into Pastoral Ministry

Our Bible School teachers were almost all former ministers or missionaries. Rev. Vernon Morrison had spent many years in Africa and had many stories to tell. He humbly told of his early struggles to preach: "I didn't get up to five minutes before I ran out of words. Then the lead pastor would quietly slip into the pulpit and finish the sermon. This went on for many months." Then one day the pastor told him, "You know I didn't think you'd make it as a preacher, till I walked by your room and heard you pray."

Brother Howells, a Welsh revivalist, would sing and have "Hallelujah breakdowns" from time to time in the middle of lessons. For the summer, teachers were expected to find preaching engagements to subsidize their teaching salaries. The maxim says "Faith is CAUGHT, more than TAUGHT." This was our "school of faith" and I would quote my teachers, especially Brother Howells for years. A visiting pastor, Marvin Forseth, showed us a pictures from a missionary visit to Argentina. "People are ready to hear the gospel" he said; "What we need are more missionaries and money to send them!"

Ellen and I had been talking for some time about going as missionaries one day. That message "sealed the deal!" We talked at length with Brother Forseth and began making long term plans. But becoming a missionary can be a long process. One of the most important items was a "SUCCESSFUL MINISTRY, leading to ordination" for starters.

Ministry Meditation:

As the years have passed I notice how often pastors make reference to their teachers' lives and sayings. Teachers are in a unique position to influence the coming generation of preachers. After much thought on the matter, I came to this conclusion; the future of an entire denomination depends on the answer to this question "What is the HIRING POLICY for Bible School teachers?"

During our days of study, full time teachers had to be ordained pastors. "Like begets like." I am deeply concerned that in our push for financial aid, government accreditation and degrees, we prefer academic rather than ministerial qualifications. Due to this emphasis, years later our Bible School had hardly any graduates going into ministry. Fortunately

Joseph G. Anonby

significant changes were made to include more PASTORS in the teaching staff.

While waiting for a church position to open, I continued my summer job of painting center lines on the highway with a six person travelling work crew. Ellen went back to being a short-order cook at Hungry Herbies. On a very busy July 1st weekend, I took an important bus trip to Cache Creek. In my suitcase I carried our engagement ring. The young Greyhound bus driver was very frustrated with heavy traffic on the old two-lane highway to Hope. "Terrible drivers" he fumed as he made a brief "rest stop" in Hope. I must have spent too much time in the bathroom, and the driver took off without a passenger count. The precious engagement ring was on the bus!

What to do? I quickly hired a young man to drive after the bus. But he needed to stop for gas. I soon realized we'd never catch up. So we stopped at a roadside phone and I called the Cache Creek Oasis Hotel, where the bus made its next stop, and told them "Take off my baggage, it has our engagement ring!" I have had lost or delayed bags on airlines over the years, but I've never been so happy to be reunited with my baggage as on that day in Cache Creek! The next day, on a hill overlooking the sleepy little town, I asked Ellen "Will you marry me?"

She said "Yes!"

While working in the Kelowna area, I got the much-awaited call from B.C. District superintendent Carman Lynn, "Will you consider taking the church in Gibsons Landing?" I prayed about it as I sat on the sand by Okanagan Lake. The Lord spoke to me by scripture, "This is the way, walk ye in it" (Isaiah 30:21). Ellen was happy to hear the news. She was planning to go into the ministry too, but when the Superintendent asked her if she'd consider going to Houston, B.C., she answered "I'm going into ministry with Joseph Anonby!" Years later, in my occasional moments of discouragement, Ellen would say "God called me to the ministry too, so you can't quit!"

Ministry Meditation:

Having a "partner" both for family and ministry is vital. You don't want a wife like Job's, who'll say, when problems come, "Curse God and die!" Problems are a part of ministry, and can lead to great victories. But make no mistake about it; we are in a warfare against evil. We need to pray for each other, cover for each other and live exemplary lives. A spouse with limited spiritual interests can destroy their spouse's ministry. If you're called to ministry, be careful in your courtship.

The church in Gibsons was a challenging place to start our ministry. A devastating church split, instigated by followers of "the Latter Rain Movement" had torn away over 2/3 of the attenders. Our congregation base was seven people over 70 and one family. A majority were "olde English folke" who loved their teas and flower gardens. But they also loved the Lord, and were so kind to give words of encouragement as they filed out every Sunday morning. Granny Wyngaert, at 83, was the eldest, and a chicken farmer's widow. She'd bring me custards, eggs and chickens to eat.

But she was clearly expecting more from her pastor. "Where is your wife? We voted for a married man!" "Have patience, Granny" I told her, "I'm getting married in October!" She obviously wanted the pastor's wife to be a good cook. "I want to see my pastor on BOTH SIDES of the pulpit!" she told Ellen very emphatically. I was the despair of Ellen for years, as she tried to fatten me up to Granny's specifications.

The church was designed and built by Saxon Sledding. Saxon had a God-given vision for building a Pentecostal church in every major B.C. Town. Over the years he built over 30 on much the same "prefab" design. The men of Vancouver's Broadway Tabernacle had the lumber transported by barge and set to work. The local priest stopped and asked, "What are you building?" He was astonished to hear them say "In three days we'll build a Pentecostal church." He probably went away shaking his head "Those crazy Pentecostals!" The church was simply-made; the sanctuary in front, and a two bedroom parsonage behind. In those days, most manses were furnished as well and were a part of the "ministry compensation." Ellen, thanked God for such a wonderful place to start ministry, and she never complained about the simple quarters.

I survived on the $80 per month salary the church was able to pay. At the time, this was the average income a labourer earned in three days. An aged Consul car served as my transport. But I knew that when Ellen joined me, we'd need more income. She worked for some weeks in a local hamburger drive-in cafe, similar to the one in Cache Creek. I helped the local chicken farmer feed his chickens. But we wanted a church "revival meeting" to get more people serving Jesus and to fill the empty benches. What to do with so few financial resources?

Ellen had an idea. "I have one more unemployment check coming to me. I'd like to donate that to sponsor the campaign." I gulped. My car was always breaking down, and the doors would often fly open as I picked up children and youth for meetings. The Rasmussens, children's evangelists with five of their own, came in their large trailer to minister. Attendance soared to nearly 100 as people flocked to hear and see the illustrated messages. On the last night, Mrs. Rasmussen stood and said "We often get love offerings, but tonight we'll take a love offering for your pastors. Did you know that Mrs. Anonby paid for this campaign out of her last unemployment check?" The people gave generously and then the evangelists gave us all their extra food. We finished these meetings with more money than we started and the church had grown.

Ministry Meditation:

"Give and it shall be given unto you" is a Biblical promise. Giving sacrificially is a form of faith. A person who wishes to be used in

serving the Lord will often lead by example. The school of faith usually begins with simple lessons we can understand, such as money matters. God trains us in these easy lessons, and when we learn, starts giving us harder lessons and greater responsibilities.

Jesus challenged "If ye have not been faithful in the unrighteous mammon (money) who will commit to your trust the TRUE RICHES?" (Luke 16:11). I contend that God will choose a generous person for greater ministry than a miserly one. In the Old Testament, fire from heaven never fell on an empty altar. You may be praying for more fire. God is looking for a greater sacrifice that He can bless.

The church folk were faithful in their work for Jesus. Mrs. Wisken had a very successful girls' group of "Missionettes." She had a simple way of making time for preparation as she also tended to a blind husband. "I tithe my time for Jesus, so I can prepare the lessons." This was a novel concept to me, and one of many examples in which the congregation taught me to be a better pastor. She rarely missed a service, and always expressed her appreciation for the messages I prepared. Another gentleman told me "Let others visit and give greetings, but you've got 'the Word'." These kind of comments were a great encouragement to me as a young pastor.

Ministry Meditation:

It is an alarming tragedy to see so many ministers quitting after very few years of service. They have, for the most part, spent several years and major financial outlays to prepare for ministry. One wonders, is the harvest of souls not less abundant if we have fewer labourers? Didn't Jesus say "The harvest truly is great, but (those who are labouring) are few...pray... the Lord to send labourers into his harvest" (Luke 10:2). But nasty comments, political infighting and scarce support cause premature "burnout" in our spiritual caregivers. Kind words, consistent loyalty and the occasional generous "extra love offering" can lift a pastor's spirit. It is also a tragedy to see churches close, because nobody is willing to pastor them. Where there are "sheep without a shepherd" the wolf will soon scatter and destroy what is left of the flock.

But one can't just preside over a little flock and make that your entire ministry. "The Lord came to seek and to save that which was lost" (Luke 19:10). We began a youth group with three Christian teenagers. It was an exciting Sunday School and youth group that brought Ellen to the Lord. She enjoyed the monthly picnics, musicals and beach parties. They would have a spiritual segment, such as testimonies and singing. But youth like to have fun! She put her youth leadership experience into practice in Gibsons. Soon we had the biggest youth group In town, with many coming from un-churched homes. Many years later, I continue to meet youth who later became firm Christians, and would say, "Thanks for that youth group, now I and my family are Christians!" Sometimes older folks would wonder about "all that fun." But we'd get them involved, especially parents of teens, and they were soon our best workers. They were our "wheels," social conveners, counsellors and promoters.

However, in a small town like Gibsons, the "church fight" and division was a hindrance to our Christian reputation. Jesus said "By this shall all men know that ye are my disciples, if ye have love one to another" (John 13:35). The faction that left were told by their new pastor, "Don't talk to the people in the church you've left!" Poor granny Wynegaert was so hurt by these slights; she didn't understand it. As we began to pray seriously about it, we felt led to approach the "rival pastor," Bro. Parks. He was not the original "splitter" and had come to faith in this independant church movement while living in Vancouver. The more we talked about joining together again, the more excited he became.

Meanwhile, Mrs. Wisken told us "I've always prayed 'God just bless them', but lately I've been praying 'God bring us together again!'" When we told her what was happening, she burst into praise. Pastor Parks told his denominational leadership what was happening. His expectation was that they would be supportive (and that we would come to worship in his new church). He was in for a major surprise.

The leader came to Gibsons, and grew steadily angrier, bursting out finally, "You're on your own, we're going to leave you!" On the basis of that vitriolic reaction, most of the congregation chose to return to their "home church" with their pastor.

Pastor Parks and I took turns preaching and teaching. The church was full and would soon be in a building program. And we were "adding to the church" ourselves, when our firstborn son, Stanley Joseph Anonby, made his appearance in time for Father's Day! I'm sure I was the proudest father present that morning, and we thanked the Lord for this wonderful child. Stanley arrived in the world via a caesarian birth. The kind doctor offered to give us a discount; there was no government medical insurance at that time. Our district superintendent, Rev. Carman Lynn graciously paid the hospital bill.

At this time we were applying for missions appointment. This turned out to be a much longer process than we'd expected. We left the Gibsons church with pastor Parks and were invited to take the work in Chemainus, on Vancouver Island. Their previous pastor had resigned abruptly due to some personal problems. We are glad to hear he has since been restored in his spiritual life.

So, we entered a church with problems once again. Yet as we drove our car with a 10 foot rental trailer into the church yard, I felt a physical sensation of the power of God come upon me. The Bible says "It is good for a man to bear the yoke in his youth" (Lamentations 3:27). With youthful optimism, resilience and enthusiasm problems don't seem overwhelming. We'd soon God work miracles in Gibsons and anticipated good things happening. One of those "good things" was the arrival of a second son, Steve Gilbert Anonby. We also received formal appointment as missionaries, though that had been delayed for awhile. I had once written a strongly-worded letter to the national treasurer against requiring enrolment in a PAOC pension plan for ordination. This was eventually made optional, but I wrote a letter of apology and have been more careful with my written words since then.

Ministry Meditation:

I will be forever grateful to the great men and women of God who have influenced my spiritual life and "made a way" for me. I call these people "gate-keepers." Barnabas, the "son of consolation" put in a good word for Paul, whose conversion from a persecutor of Christians to a foremost apostle was suspect. I advise our Bible School teachers and future pastors not to "squash" dissenters. They are potential future leaders, who have new ideas or need to be assured that we are doing things in the best way possible. At the same time, David prays "set a watch before my mouth" (Psalm 141:3). You can imperil your own future by "putting your foot in your mouth." It's not wise to belittle your leaders – you may become one yourself. When unwise words or actions occur, repent quickly; "the best time to eat crow is when it's still hot."

We ministered to the best of our ability in this company sawmill town of around 3000. The church was still shattered by their previous pastor's conduct. We prayed for a "Holy Ghost idea" of what God's plan was for these dear folk. A previous moral failure within the congregation brought further division. Many simply wanted to quit and worship elsewhere. At the same time, a native congregation under the Millers had lost their place of worship on the nearby reserve. We were able to unite these congregations as a single body of believers, praise be to God.

"Ordination is required for appointment to full time missionary work." Ellen and I went through the interviews with the executive and were approved along with four other of my colleagues. The standard procedure of the time was to repeat the vows of ministry and then be prayed over by the presbytery. Many times there was a special prophetic word, and that night there was a word over each candidate.

The "word" for me was this: "Remember that the power and the wisdom you have comes from God." I have never forgotten those words. A dear old saint, Miss Herd, from Chemainus was there that night. She saw a dramatic visible manifestation of the Holy Spirit's presence as I was being prayed for, but I don't feel at liberty to reveal those details at this time. Suffice to say that we were greatly encouraged that God was with us. We prepared ourselves to go to the mission field.

Chapter 3

Preparing for the Mission Field

"Itineration" is not an official word in any dictionary I've consulted, though "deputation" is often used by many missions organizations. We contacted several of our pastor friends on Vancouver Island and preached about our call to Argentina. God had prepared our hearts to expect miracles of open doors and His provision. One outstanding example of God's provision happened shortly after our arrival in Chemainus. Prospective missionaries were expected to attend the School of Missions in Peterborough, Ontario at their own expense. IF they were later appointed officially, costs would be refunded. That's many days of driving from B.C., and we had only $20 two weeks before leaving. I determined to not borrow money for the trip. If God wanted me there, and He wanted us to be missionaries, we prayed that He'd provide. This is how God did it...

1. We were in the middle of a multi-car collision while bound for my brother David's wedding. I got paid nearly $165, very quickly and much to the surprise of agents. It cost me almost nothing to repair only what I considered necessary.

2. Mother told me "Our pastor got church support while he travelled to the School of Missions." I asked the board and received a partial salary of $35 per week while travelling.

3. The promotional missions secretary arranged for a series of meetings for me in Ontario.

4. My brother Daniel gave me money for a new tire.

5. We bought a little tent and camped along the road for the week of travel.

6. An Ontario pastor, Maude Ellis, lent us her summer cottage at Cobourg between meetings.

7. We ran out of gas once, and partially filled the tank with our Colemen stove fuel. Then a telephone serviceman in a rural area filled our tank from a company depot.

Joseph G. Anonby

8. We met pastoral friends from B.C. in the Winnipeg national PAOC conference and they gave us an additional donation.

9. Other friends gave us a free place to stay.

10. We still had to pay for the ferry trip back to Vancouver Island, plus we needed a new battery. I paid for a $20 battery on a credit card.

11. We came back to Chemainus, with $20 in cash left over—what we'd started with six weeks earlier.

We continued to itinerate in B.C., and visit mission supporters. But the entry visa for Argentina was very slow in coming. Rev. Lynn, now our national missions director, suggested we move to Eastern Ontario, to be closer to the Argentine embassy in Montreal. We continued to visit many churches, as we awaited the visa. It was a frustrating time, but the Lord gave me a series of songs that encouraged my soul. One of them went like this:

"All things work together for good to them that love the Lord (2x)

He said it in his Word, we know God's Word is true,

All things work together for good to them that love the Lord"

Years later, our children would scold me if I got too uptight, "Remember Daddy...'all things work together for good to them that love the Lord.'"

Once Ellen and I were praying, and wondering what we should do as time dragged on. Suddenly the Spirit of God came upon me, and I prophesied a Word from the Lord that went something like this: "I have called you and I will open the doors. Trust me, I will lead you and provide for you. Do not doubt what I am doing. I love you and will care for you." It was a miraculous moment. Never, before or since have we experienced such a direct word just to us two.

We were filling a pastoral vacancy in Whitby, Ontario at the time. Only a few weeks before, the Lord had begun to entrust me with words of prophecy, tongues and interpretation. I began to recognize a feeling of pressure in my spirit, and the first sentence or so of God's message. I stepped out in obedience, like walking on water, and God gave further sentences as I continued. That's how I've experienced God's Spirit moving in me to minister words from the Lord. At times a holy anointing in similar fashion occurs as you preach, and a torrent of words pours out. These may include words of wisdom or knowledge, "the self-same Spirit, dividing to every man severally as He will" (I Cor. 11:11).

Ministry Meditation:

There are always elements of mystery about the things of God that we'll never fully

understand. But God does give us some guidelines in the Word. "Desire the best gifts, but rather that ye may prophesy...to edify the church" (I Cor. 14:1, 4). "Desire" indicates some measure of spiritual ambition on our part. Paul said "I press toward the mark for the prize of the HIGH CALLING of God in Christ Jesus" (Phil. 3:14). Phillipians 3 is my "life motto" chapter. I had been praying many years for greater spiritual gifts or "charismas" to better serve the Lord. Vocal spiritual gifts were a common weekly occurrence in my early church experience in Nelson. We are commanded, "Quench not the Spirit" (I Thess. 5:19).

Quenching the Spirit or "forbidding to speak with tongues" (I Cor. 14:39) is a serious offence. It is putting human rules over Divine authority. I believe pastors who require many human rules for the exercise of gifts are themselves "out of order". Sometimes critics silence gifted people. I recall the sad tale of a fellow-missionary who began to fear giving messages in tongues in a church with several non-Pentecostal adherents. "And after that" she sighed "God never used me again."

I have a personal concern about "worship leaders" who act more like concert-performers, singing their list of songs. They fail to recognize when the Spirit is moving, and give Him no place. Their time is spent more in practice than prayer. The Lord often moves in times of contemplative silence after a solemn worship song. Frequently I have been sorrowful as I felt the familiar move of the Spirit whelming up in me. A message was about to burst out; the first sentence of utterance was on my lips. But the singers just start another song without a pause. I believe some pastors may have abdicated their place of spiritual authority over worship leaders. We are told "Grieve not the Holy Spirit" (Ephesians 4:30). Over the years I've seen a steady diminishing of the active use of Spiritual gifts. The work of God is poorer as a result.

But God's supernatural guidance is not usually a conscious daily occurrence. Among God's giftings for the church are "gifts of governments." Our missionary Director, Rev. Lynn, wrote, "When we keep knocking at a door for some time, and it doesn't open, perhaps God is showing us it is time to change course." I'd talked about further education and he suggested applying to the Newfoundland Assemblies, who had their own government-funded school system at that time. I was accepted by them on the basis of my Bible School graduation diploma.

Still, there was some hesitation about going to Newfoundland, the last British colony to join Canada as a province in 1949. My eldest brother, John, was going to get married in Edmonton, where he was studying for his Doctorate in English literature, with the help of a prestigious Canada Council scholarship. He helped pay some of our car travel expenses to Edmonton. My old Chev had "died" and friends had given us another Chev with "bat wings", year 1963 I believe. It was rusted out, as many Ontario cars were due to the salted roads and freezing rains. One day we drove over 1100 miles, and came to Edmonton in short order.

John had been a bachelor for many years, so I firmed up his resolve to get married; ("Yes, of course you'll lose your freedom, but it's worth it"). He was also habitually late. The night before the wedding, I borrowed everyone's watch and set them a half hour ahead. We were heading for the church when John had yet another task to do, "But there's no time!" I told him "Go ahead, relax, we're a half hour ahead of time." He'd forgotten the rings too, so "with this ring" (MINE and the pastor's) "I thee do wed."

From Edmonton we drove to Castlegar, British Columbia, where Ellen's parents had returned to live. The new two year Selkirk College had just been built there, and they were eager for students. I qualified for immediate entry having grown up in the area. In short order God arranged for a scholarship for me. Pastor Simmonds needed more room and bought his own home leaving the church parsonage vacant with a modest rent of $25 a month. We got a store-cleaning business going in very short order that covered our expenses by working one night a week. Soon we were leading in Sunday School and youth services. The only sour note was the Superintendent's notice in the B.C. Pentecostal Newsletter: "Joseph Anonby is no longer serving with the missions department and is going to a SECULAR college."

But I had not given up on becoming a missionary. Our visa application had been turned down by the Argentine government. We later learned that a group of 200 Mormons had falsified their residence applications to pose as "book salesmen". The government threw out the whole batch of applicants, and our application was thrown in with these rejects. Sigh...

Still, I thought, "If I can't come in as a missionary, maybe I can gain entry as a teacher or psychologist." Meanwhile, we wanted to continue in pastoral ministry. We took my Dad down to the Coast for a TB test, and talked with the Superintendent about our dreams of both pastoring and studying. "Impossible" he said, (although his own son had done that in Quebec), "there are no churches available for that; they are all larger." With those words ringing in our ears, we drove the nine hour trip home to Castlegar. All the way we talked and prayed about it. When we arrived, we told pastor Simmonds, "God is calling us to Vancouver, and we believe He'll open a church for us."

We sold our cleaning business to a nearby pastor, Rev. Charles Postal, in trade for his car. Our "rust bucket" had dangerous holes in the floor boards. On our first Sunday in Vancouver we went to a little church at "Capital Hill" N. Burnaby, where Ellen used to attend while at Bible School. Their pastor was leaving and it looked as if the church would close. "Please, please be our pastors!" they pleaded. They had an upstairs parsonage ready to move into, and move in we did. Simon Fraser University, recently-opened, had a trimester program available and was looking for students. It was nearby. The Department of Education gave me several bursaries and I earned some scholarships to help pay for the psychology and education studies. Our daughter Joy was arriving soon, and we needed all the help we could get. Still, by living frugally and careful borrowing, we were able to buy our own furniture, for the first time in our lives. No "parsonage furniture" was available in this church.

Ministry Meditation:

I well recall the day Ellen and I were still dating, but thinking seriously of a life together. As we walked by some new homes near the Bible School, she said, "I don't ever expect to live in a house like this. Serving the Lord and living for eternal values is the most important thing in life." That took a lot of pressure off me. If you are a "status seeker," certain appearances such as your neighbourhood, profession and income are indicators that you are among the elite (or not). A minister has to "count the cost" and decide whose approval he is seeking. If you're looking to be wealthy, ministry (with some controversial exceptions) is not the place for you. It is vital that BOTH husband and wife share the eternal values of seeking God's final "well done" rather than public prestige. Beware of a wife (or husband) who is always wanting MORE things.

I continued my studies at SFU, and enjoyed being a part of the Varsity Christian Fellowship. The church began to grow and we were considering expanding the facilities. Suddenly, the long-awaited visa to Argentina arrived. The government immigration agent in Argentina told Superintendent Grasso "Of those 200 rejected ONLY THIS MAN GOES!" But...we had to make a new application for our baby Joy. That took a whole year to process! Some of the new converts and attenders were discouraged to learn we'd be leaving. Well, we HOPED to be leaving soon...but there was the matter of some debts.

On our first application, we also owed some money for a car. The Lord helped us sell a cleaning business to clear those debts. Now we owed money for furniture that the company was not interested in buying back. We took a service in a nearby church on Boundary Road. About a year earlier, we'd paid for the car repair of a young couple from there who'd been serving in Youth with a Mission. As I gave the money to the young fellow, he said "But you can't afford this!" I replied, "I know, take it fast before I change my mind!"

When I preached that morning, the Lord moved in a marvellous way in this church which had once rejected spiritual manifestations. The Lord gave me an extended prophetic word that in essence said the following. "God wants to bless his people, but they need to be obedient to the Word, and be generous. The children of Israel grew miserly with their gifts to the temple, so the priests could no longer teach the Word and offer sacrifices. Their children, now untaught, fell into sin and brought down God's judgment. They were sent into years of captivity and were impoverished because of their disobedience in giving to God." I trembled as I gave this word under a strong anointing.

The reaction was swift. Pastor Carlson jumped to his feet and shouted,"I've been faithful in my tithes but we only gave $12 to missions last month." He turned to his wife, "Katie, write out a cheque for $50 to missions." The young fellow whose car bill we'd helped pay exclaimed "This is a man who practices what he preaches." The missions offering that morning came to over $900. We paid our debt and wrote on the missions re-application "Yes, we are free of debt!" Years later, when we needed new missionary support after leaving Barcelona, we got a $10,000 donation from this couple whose car bill we'd helped pay years before. "Give and it shall be given you, good measure..."

We did some church itineration once again, but this was brief. We had to be in Argentina by October, or the newly-issued visa (with the addition of Joy), would lapse. There were happy goodbyes to my parents who'd always loved missionaries, and sad goodbyes to Ellen's folks "Why are you leaving me to go so far away?" I was not very popular on that side of the family.

Chapter 4

Living in Argentina, our Mission Field

We arrived in Buenos Aires, Argentina, exhausted after a 20 hour plane trip. On the N. American part of the trip we were reprimanded,"get those children out of the aisle!" On the S. American segment the stewardesses kept running their fingers through our children's blond hair, exclaiming "Precious, marvellous kids!" We were ushered immediately to the front of the immigration line with our tired children.

Former Bible School friends, the Frickers and Nerlings, had already arrived as missionaries. Werner and Heidi Kniesel, missionaries affiliated with the German branch of the PAOC, welcomed us into their home. Villa Ballester was an area of Buenos Aires largely settled by German immigrants. I'd studied some Spanish at SFU, but when I stumbled in my Spanish, people would see my blue eyes and immediately speak in German. I learned to say "Nicht verstehen, hable Espanol." ("I don't understand, talk Spanish"). Soon we were settled in a former doctor's home, and began our year of language study "on site," as was PAOC policy.

We made our share of classic language mistakes as did our friends. Missionary Fricker was helping me look for a bed, "Just simple, and cheap" he'd ask, and time and again we were shown only expensive beds. I was sick with what turned out to be a measles relapse. At one store he replied "Too cheap." Even with my rudimentary Spanish, I knew he'd goofed. As we left, he mumbled out the side of his mouth, "should have said 'too expensive'." I completely lost it and dissolved in laughter as we exited. The owner looked at us oddly, "Crazy gringos!"

Another time, Ellen was shopping and asked for "500 kilos of cheese and 700 kilos of meat." They would smile and give us the amount in grams, instead of over 1000 pounds of meat and cheese. She also told a "street-parking" official "I'll hit you later." Oops, she meant to say "I'll pay you later." People were very kind, and the youth especially patient, since they wanted to learn English. Not long after our arrival, I was asked to lead in prayer in the local church. Prayer in Spanish requires the extensive use of subjunctives for wishes such as in "correct English" you say "I wish I WERE" (NOT "WAS")." It was a VERY LONG time before they asked me to pray in public again!

Our children had been worried about learning Spanish, "You are studying, what about us?"

"Oh" I replied "Children learn languages like magic!" We were very pleased how they could soon communicate with our neighbour's children. But some guests were horrified. "Do you know what they are saying?" Umm, no. Oops, they had been mastering some earthy swear words.

"To learn Spanish you need to drink 'mate'" we were told. This is a herbal tea that has a stimulant like caffeine, and is served in a gourd. It is commonly drunk twice a day. You sup it via a steel straw with a sieve at the bottom. It is routinely passed from one person to another. Sugar, lemon or more "yerba mate" is added from time to time. The straw is not wiped clean between servings.

Stan was getting a lot of colds, mostly due to insufficient heating in the tiled-floor house. We got gas from underground street lines, but during high demand winter days, there was so little gas that even the gas pilot lights went out. The doctor told us, "The boy needs to have his tonsils removed. Bring some towels to the clinic." The "operating room" was a reclining dentist's chair. With a covering sheet and a minimum of anaesthetic the doctor snipped off the tonsils. As the blood gushed out he snapped "Give me the towels!" We stayed by Stan's bed as he thrashed and cried during a restless night. For many months he had a phobia for white uniforms or anything that covered his head and eyes.

We kept getting "culture shock." For example, I'm used to being punctual, and became critical of late arrivals and "no shows." The Lord taught me a lesson quite quickly. I'd been asked to fill in for the pastor, Angel Furlan, while he went on vacation. There was a British lady who would translate. The problem was–I never showed up! At meeting after meeting the urban electric train would malfunction due to downpours. Taxis couldn't find the church. When I finally made it, the translator said "When you didn't show up, we asked a visiting preacher to speak, and several people were saved. Probably wouldn't have happened if you spoke!" Thus the Lord keeps us humble. And I often said "The Lord kept me waiting over three years for my visa to teach me patience in the Argentine culture."

I'd done some research about the country. Argentina is the 14th largest country in the world in terms of area, and in 1970 was the 20th in population. It was a nation of immigrants from Spain, Italy, Germany and Russia, in that order of numbers, with about 30 million people. The native Indians, many living on the large prairies "pampas" were displaced by settlers riding wagons similar to our "prairie schooners" large two-wheeled carts. No treaties were made with the natives, nor were they given reservations to live in. Military leaders took charge as heroes, and to this day bloody victories over the Indians are celebrated yearly.

"Martin Fierro" is an Argentine epic poem written in the mid 1800's. It describes the daily life, travels, songs, philosophy and woes of an Argentine "gaucho" (cowboy). There are over 60 million sheep and 40 million cattle on the pampas (prairies) of Argentina, so many folk have rural roots. The gaucho in this classic poem rejoices in the love of the land and sleeping under the stars during annual round-ups. Yet he is often a social outcast, abused by his employers, unlucky in love and attacked by Indians. Many times he has an Indian mother.

But he has his pride and is itching for a fight with anyone who offends him, however trivial the reason. At the end of the poem, Fierro intones lines which ruefully admit that pride is the primary cause of his troubles. Short phrases and couplets from this classic poem are a part of Argentine discourse and philosophy to this day. Like Paul, on Athens' hill, I found that quoting "Martin Fierro" (even in sermons), would always bring a smile.

"How do you start a church?" was the burning question on my lips. I'd always been interested in Bible School teaching and thought that might be my initial ministry. Not so fast! A member of the national executive which interviewed me asked nervously "Have you been told you were to teach by those who sent you?" (A former priest, he was anxious to retain his position as Bible School Director). The consensus was "Go start a church first!"

We set ourselves the task, while learning the language, to also educate ourselves in being church pioneers. Here was our strategy; invite as many people with these skills to our house as possible and "pick their brains." Ellen made a nice supper, and afterwards I'd take out my notebook to ask, again and again "How do you start a church?" The national leadership had targeted nearly a half of the 24 provincial and territorial capital cities that did not have an Assemblies of God church. These were cities far from the huge metropolis of Buenos Aires, which, with close to 10 million people, contained almost half the country's population. National workers were often reluctant to go far from home to pioneer a new church. But missionaries? Well, they'd already left home far behind...

Our chief mentor was Paul Brannon, a clever and straight-spoken missionary-evangelist from Kentucky. He'd helped other missionaries hold campaigns and start new churches. We went together to northern Argentina, passing province after province with no Assemblies of God work, not even in their capital cities. At the end of a three day journey, we crested a hill to look down on Salta, the northernmost province, with a capital of the same name.

As I looked over the city, I felt the physical sensation of a mantle of God's anointing come over me–the same experience that occurred in Chemainus, years earlier. This was the place to start our first church! The city lay on a flat plain near the Andes mountains to the west. It had a population of 190,000 in 1971, and was growing by 10,000 people a year. Sugar cane plantations, a large tobacco industry and ranching were the main sources of employment.

After 18 months of pioneering, we had a church building on the main street and a congregation of about 40, praise the Lord! How do you start a church? Here are some hints of how we did it in Argentina.

1. Get God's mind as to WHERE the Lord is calling you. That will hold you steady during testing times.

2. Educate yourself by listening to mentors. This includes researching books and articles.

3. Talk to pastors in the area. They may not all be very welcoming. (One missionary's wife told me bluntly "We have all the churches we need here already." I shot back with "That's what they said when YOU CAME too, didn't they?") They went on to begin what became the largest church in the city…

4. Get a group of Bible School youth to be your "start-up team." Some stayed for ten months, living in our crowded home. That's a "culture-adaptation experience!"

5. Advertise extensively: posters, radio spots, pamphlets, door-to-door invitations, rapid street meetings/announcements. God may be moving, but people need information to attend.

6. Literature for salvation and follow-up with new converts. Note addresses for visitation.

7. Rent land for meetings, using a tent if possible. You'll also need a caretaker.

8. Evangelists: line up a list of capable people familiar with public ministry to the un-churched. HEALING MINISTRIES are very effective.

9. Hold teaching sessions for new converts, preparing them for baptism, membership and the infilling of the Holy Spirit.

10. Nightly meetings for several months. Have good Christian films and music groups.

11. Picnics and other social events give a "family feel" to the church.

12. Start specialized meetings for youth, children, ladies and men.

13. Begin looking for a permanent church building, and solicit offerings for that goal.

14. Have a least one helper, (a national in this case), who will become the pastor.

15. As a missionary, be prepared to "move on" and let a national be the pastor.

16. By God's grace, in 2013 we started church number 14, in Santo Domingo, Dominican Republic. Every church has a different story of how it was founded. But some of these basic principles are always present. It is clearly God's will to save souls, and He said "I will build my church." It's a thrill to be a partner with the Lord, winning the world for Christ in obedience to the "Great Commission."

Ministry Meditation:

Point #15, "move on" is very hard to do for some missionaries who have built a thriving congregation. But "mega-churches" often fall flat when the founder leaves. Without a constant injection of money, programs begin to stall. If a gifted national takes charge, he'll continue to grow the church with occasional missionary input. Control conflicts are avoided, as are major financial disasters for the national church.

An additional observation; it is better to leave voluntarily, than to be "pushed out!" I know several good missionaries whose careers have ended prematurely because they would not transfer a church, Bible School or other ministry into national hands. Yes, nationals will make mistakes, as we all do, but God will guide them. Paul left his newly-founded churches in the hands of capable elders and went on to start other churches "in the regions beyond." He then sent letters and helpers to continue guiding "his children in the Lord."

The Lord did some unusual things as we began the church in Salta. We rented a lot beside the well-known "Plaza Gurruchaga" and had meetings every night. Word began to spread "People are getting healed, and alcoholics are kicking their addiction." Good news spreads quickly, and missionary Ralph Hiatt donated some months of radio time for a church program. He also painted rapid chalk pictures that were given to the person who brought most visitors. He once preached, amid the frequent rains, "We came to this MUD to preach the gospel!" He meant to say "THIS BARRIO" (neighbourhood), but said instead "BARRO" (mud).

A sizable group of students began attending. One that stood out (very tall), was Willy Grimolizzi, one of nine children whose mother owned a drug store. He arrived with his girlfriend, prepared to mock and make a public scene as he'd done in previous evangelical meetings. But God got a hold of Willy that night. I prayed with him for salvation as he sat on our crude benches. When I asked "Do you have some other needs?" he replied, to my surprise, in English "I want to get off drugs!" That night he threw his pills onto the sawdust floor and walked out a changed person.

National pastor Jose Vena encouraged him to give his testimony on our radio program. The director of the technical/electrical school where he was taking studies was very upset. His own mother said "I'd rather you were on drugs than become an evangelical." But Willy went on to Bible School, along with three other recent converts. He started a new church with missionary Parks in the provincial capital of Santa Fe province. And now he is his mother's "favourite son!"

Carmen had emotional problems, and a phobia about her "ugly legs." She often would hide her legs under long pants, retreat from the public, and cry. But God got a hold of her in a camp meeting we organized for the new church. She was filled with the Spirit and new boldness. "Look at my legs!"

she exclaimed as she swirled before me in her new dress. "Quite nice!" was all I could say. Another convert was Alicia, our neighbour's maid. She'd been orphaned as a child and raised with abusive relatives. For a while she attended a Baptist church, but slipped away and joined her employer in smoking. Soon she used her voice, still husky from smoking, to sing praises to God. We had "middle class" and "low class" converts all enjoying equality in Christ. Carmen went on to start a Bible School in Salta, and Alicia became a pastor's wife. Ernesto Somarriva, a former rock band musician, became a church-founding pastor in Buenos Aires.

God had a special healing miracle for our middle son, Steve. He loved chocolates and cocoa, but was allergic to them. When evangelists prayed for healings, Steve would pray too. The first time he was disappointed, and we cautioned "Sometimes God says 'No'." He mulled that over, and the next healing service went forward again. As soon as he came home, he poured himself some chocolate milk and declared "This time the Lord said 'Yes'!" It was true. His chocolate allergy was gone.

Time to move out of the tent, into a permanent building. It can get cold in Salta, not far from the towering Andes. On the main street (named "San Martin",– the "liberator" from Spain), there was a large building that once been a mattress factory. It now served as a candle factory, directed by Belloni, a humorous Italian. He joked "I sell candles made of cow fat wax. Every year they have a big Saint's Day in a shrine to the south. Afterwards, I drive down and recycle all the wax for next year!" We were able to get the building for about $15,000 and paid for it over three years. With a failing peso, it ended up costing us under $13,000.

This was enough money to start building on a parsonage at the back of the lot. One of the new converts had an old father who built cement-block houses. About half-ways through, he had a stroke, and we needed a new contractor. On the second day, as he was laying parque (wood tiles) on the floor he asked "Could you come to look at this? The room is not square." Would you believe it, the house was one meter shorter on the south-west side? Ever after it was known as "the house that Anonby built!"

Ministry Meditation:

It was our goal, ever after, to provide a parsonage for the national pastor. Many times their entire salary would be needed just to rent a decent house. Unless the church was able to sustain them fully, they'd need to work full time and have little strength left for ministry. Even in North America, when pastors change, or the church loses members, having a parsonage is an added incentive for a new pastor and a saving for a church with limited resources. Churches with a home for a pastor rarely close permanently. Of course a larger church is able to give a housing allowance so pastors can choose to rent or buy. Buying in a small town is not always the best option, but if a pastor can eventually buy his own home, he won't dread the day he has to retire.

Anonby Adventures in Missions

Just before we officially bought the church building, we had a wedding in the tent. The place was packed with people curious to see what a Protestant wedding was like. My five year old son, Steve, was the ring-bearer, with his white shirt-tail sticking out. We had an abbreviated service before the wedding and received an offering with more one peso coins than I'd ever seen. We got permission to use the new church for a reception. Jose Vena, the national pastor, persuaded several soda drink companies to donate refreshments. Ellen baked some cookies, but there was a walk of at least five city blocks from the tent to the church. Surely not everyone would walk over. Oh yes, they did! What to do? We passed out free drinks over and over again. Even the ravenous children groaned "Enough!" Then we passed out the cookies...and there was enough for all the (limited) appetites. This was our Argentine version of the "feeding of the 300!"

Chapter 5

Starting a New Church in Jujuy 1973-74

The church in Salta was now set in order, with deacons and a baptized membership. It was time to apply the above-stated rule #15 to **"move on."** We considered starting another church in a new area, the "barrio Tribuno" with 5000 inhabitants. But our Canadian missions secretary, Rev. Lynn, thought it best to go to another capital city, Jujuy, in the bordering province of Jujuy. This bustling city of about 90,000 had a large percentage of Indians (Quechua and Aymara). Often small and stout, many were descendants of the original inhabitants and others from La Quiaca, Bolivia, just across the border at 9000 feet in elevation. There were no other Assemblies of God churches in this province.

We began once again in a tent, with Carlos as the live-in caretaker. Meetings were again held every night till a faithful group of new Christians was formed. Maude Ellis, an evangelist from

Ontario, Canada, prayed for the converts in both Salta and Jujuy to receive the baptism in the Holy Spirit. And how the Spirit began to fall! Our three children, Stan, Steve and Joy, ranging in ages nine to five were among those blessed with a heavenly language.

At the same time, there were some new ministerial blessings we had not anticipated. Ellen climbed over benches to lay hands on seekers and translate as Sis Ellis told them "Stop speaking in Spanish, the Holy Spirit is upon you, give your tongue to the Lord and launch out in God's language." We soon after received the same mantle of faith to encourage people through to the baptism. It has been an important part of our ministry ever since.

Ministry Meditation:

It is my observation that ministries are "contagious." In part, it comes from being mentored to become a successor. We see that with famous "pairs" such as Moses/Joshua, Elijah/Elisha, Paul/Timothy.

Many ministers of renown have children who follow in their footsteps and eventually assume their positions. But the vital factor is that we are eager to learn and "covet the best gifts." Stay in God's presence, like Isaiah in the temple till God calls out "Whom shall we send?", and you can reply "Here am I, send me!" While it is God who determines whom He will use, we can greatly increase our chance of being chosen if we lead an active, pure and prayerful life. Add to your Bible knowledge by a good theological education and you will avoid doctrinal excesses which can derail a successful ministry.

We were thrilled that four of the new converts from Salta, our first church start, were going to Bible School. These were the most promising youth recently-converted and excited about learning more about the Bible. The difference between a sincere believer and a Bible School graduate in terms of effective ministry is often dramatic. They begin to learn pastoral skills and feel the joy of God using them. Denominational leaders and individual churches are more confident to employ a trustworthy pastor. But you can understand the mixed emotions of our national pastor. He would sorely miss the help of these new converts.

Ministry Meditation:

It's not easy for a pastor to "give up" his leading youth converts to go to Bible School. They may never return. In fact it is quite common for overseas national pastors to deny their applications or demand they return for a year or more to help in the local church. I've observed two negative results of this attitude of "ownership." In my home church in Nelson, my father told me "A Pastor didn't want Dolores to go to Bible School, since there were few youth." But Dolores married a non-Christian who later got into problems with the law. She continued attending church, looking sad all the time. A Bible-School trained youth is eager to work for the Lord. But their ideas can become a challenge to insecure local pastors, who then "put them in discipline" (quite common in South American churches). Gradually, their skills and zeal for the Lord fade, and the Lord's work loses a labourer.

We need to celebrate and encourage those who wish to go into ministry. Pastors who facilitate workers become participants in their work. They can also help sponsor students, who then give to the local church and missions, or work in all kinds of "helping" tasks. They are all "workers together with God." In hockey those who pass the puck to "assist a goal" also gain points. God is keeping track of all phases of His work. "He...that goeth down to battle...and he that tarrieth by the stuff; they shall part alike" (I Samuel 30:24).

In Jujuy, our first order of business was to find a place to live. A recently-built large steel plant nearby had brought an influx of engineers who rented most of the available houses. The only option was renting a "summer home" in the nearby country town of Yala. We were on the edge of town beside a large river. The house roof was made of cement/asbestos sheets, with a few gaping holes. There was a small and leaking "fill and dump" pool in the back yard. A large fireplace gave some heat when we could drag wood up from the riverside. It was primitive, but our children loved it.

In the winter there were so many ticks in the dry grass that our children had to stay on the porch. The country had imported insufficient natural gas for the winter, and that year, gas prices spiked. I spent a large part of the week hunting for bottled gas as we gradually ran out of gas for heating, washing and showers. We only had one gas bottle left for cooking. To keep warm we wore many clothes, fur-lined boots, gloves and heated our hands over morning hot chocolate. One visiting missionary family slept three to a bed to keep warm. But our mostly-poor new converts exclaimed "What a palace!" That spoke volumes about their living conditions.

In order to keep down out-of-control inflation, the government would place price controls from time to time on the basic "family food basket." The logical result of that, after awhile, was a shortage of these articles. One embarrassing shortage was toilet paper. We were down to the "critical stage" when I went downtown one morning. I was delighted to see the streets littered with a recent "advertising offer" dropped by a small plane. Quickly parking my car in this busy zone, I scrambled to pick up paper. Someone joined me, but stopped abruptly when I said "This will be our T.P!" A policeman scolded me for my ill-parked car, but gave me a pass when I pleaded, "Sir, I need some T.P.!" They weren't quite the Sears catalogues of olden days, but they did the job.

Starting a church is a very time-consuming task. There were so many things to do that I sometimes forgot important things. We were almost at our pioneer "gospel tent" one Sunday, when I noticed "Oops, I forgot my Bible!" It had my Sunday morning sermon notes that I'd prepared earlier that week. Would you believe it? I couldn't remember what the sermon was, nor what the Bible portion was! Ellen and I prayed **fervently** for inspiration and now I stood behind the pulpit to preach. I opened the (borrowed) Bible at random to Ecclesiastes 3 **"to everything there is a time". That was exactly the message I'd left at home!"** All the main points came back to me during this totally unplanned "extemporaneous preaching." Thank you Jesus!

"Bees, everyone into the house!" I yelled one day. A large swarm of **African bees landed on a banana tree in our yard.** Some years earlier, during an experiment in Brazil, a hive of deadly African bees escaped from the bee-keepers. They are hard-working bees which produce no honey, and drive out the local bees. Over the years they have moved about 200 miles each year in both southern and northern directions. When they sting, they release an odour that excites the swarm to attack. Many people die every year from their attacks. Steve was especially allergic to bee venom. After a few minutes they left the banana tree and began entering our bedroom through a space in the shutters. We shut the bedroom door while scores of bees congregated on the ceiling. I got a stepladder and a can of "Raid", then hurriedly climbed up and sprayed them. I did this several times, and shut the door again. After about a half hour we entered and found hundreds of deadly African bees on the white cheneel bedspread. Our house was attacked twice, and we thank praying people for God's protection.

Once Stan was climbing the rocky slopes above our house on the outskirts of Salta. He was bit severely by something, but couldn't find the wily culprit. At the hospital the doctor said "It's a snake bite! I'm an expert in this. It's lucky you were only pierced by one fang!" He was hospitalized for awhile. Conditions were appalling; there were feces on the floor. Steve said "Let's get Stan out of here!" We were travelling in evangelism on a "goodbye" tour at the time. We prayed, took along some pills, and continued ministering in the next place. Stan recovered quickly, ptl.

Ministry Meditation:

People who have not lived overseas in long-term missions work sometimes have unrealistic ideas of how missionaries ought to live. Some feel they ought to live "very poor." I know people who have done just that. But that is not recommended by experienced missions organizations. One reason is preserving health and physical safety. Living beside putrid water and street cesspools or a crime-ridden area can shorten your missions career PERMANENTLY. Also, such missionaries don't usually continue as lifetime workers. In addition, many nationals have very strong immune systems, and do not expect missionaries to be as poor as they are. They would prefer to have someone to look up to, and are looking to improve their lives. A missionary is expected to provide employment and maybe help with school fees. A very poor missionary is not someone they are proud to know and recommend their church to others.

But the opposite extreme can also be a problem. In Paraguay, missionaries were able to buy a large and inexpensive tract of land outside Asuncion, the capital. A missionary built several American-style houses surrounded by fruit trees. The President's son attended their Christian school. However in Paraguay, about half the people go barefoot, so the contrast was stark.

After some years of difficult missionary work, and comments from the national superintendent, the houses were sold, and replaced with simpler ones. Missionaries are looked at closely by nationals and visiting ministries. We need to be careful of "conspicuous consumption" while some supporters are living below their means to give to missions. But if a missionary is generous with their possessions, and hospitable, it reduces envy. And things do change. Today, Asuncion has one of the largest Assemblies of God churches in South America, with many upper-class people living in nice houses.

We normally had a potential pastor in the group of students that helped us begin the church.

New converts get to know the person who was introduced as their future national pastor. The missionary assumes a supporting role, and his experience helps in the training and grooming of the national. But in Jujuy, we were not able to get a suitable pastor candidate among the student team. Some were new converts from Salta. What to do? Evangelist David Peloso came to minister in our tent. Simple and powerful in his preaching, he won a place in our hearts. "Where can we find a pastor before we return home on furlough?" we wondered. He offered "I'm looking for a place up north to pastor!"

We had been looking for a building to transform into a church. It was a full time job for me, and still I found nothing suitable. Finally I told the student group, "Do nothing else but look for a place to build a church." A few days later, they asked a taxi driver who was washing his car "Do you know of a building for sale around here?" "No" he replied. They shot back with "We'll buy your house!" And we did! Pastor Peloso built a two storey house in the back yard, and transformed the taxi-driver's house into a church. The Pelosos, with their daughter Lilian, lived with us while he did this construction. We stayed up late talking and laughing at David's many stories. Mrs. Peloso was never able to finish a funny story; she'd laugh so much that her husband had to give the punch line.

Their daughter, Lilian, was still alive by a miracle. Some years earlier, their pastor had challenged his congregation to "promise to pray one hour every day." David promised, but kept falling asleep. "What are you doing taking a shower, a COLD SHOWER, at midnight?" his wife protested. "I promised to pray an hour, and I'm going to pray one hour!" he replied. As he returned to prayer, God's Spirit moved in and he prayed for something like 15 hours straight. He arose, a changed man and a powerful evangelist. Like most Argentine evangelists, he prayed for the sick. But he soon faced his biggest challenge yet.

As he prepared to go to a meeting, his daughter was taking a shower in the gas-heated bathroom. "What's taking her so long?" he complained. His wife went in to check and came out screaming, "She's dead! She fell in the shower! She's not breathing!" The heater was leaking gas fumes which knocked her out. David's brother came over, while David began praying. "Get the doctor, don't just pray!" he shouted, and ran to get the police. David kept praying, while people consoled his wife with "You can have another baby." The police showed up demanding "Show us the dead girl!" Just then Lilian came to, having regurgitated from gas-fumed lungs and stomach. "Here's the girl alive!" announced David, "But there are people here that are 'dead in trespasses and sins'!" He used the opportunity to preach the gospel. The police glared at his brother and jailed him for giving a fraudulent accusation. Ironically, David had to bail out his brother!

Long nights of meetings that lasted for months in both Salta and Jujuy began to take their toll on my body. Maude Ellis became concerned when she heard my frequent long sighs. "Joseph, you're getting overtired, a doctor once told me that long sighs spell exhaustion" she warned. I thought I had my recurrent bronchitis, and was self-medicating with antibiotics. Finally I noted my 'water' was getting very dark, and the doctor diagnosed "It's hepatitis!" It was severe enough that he sent me to the hospital for ten days. Stan also came down with the same disease, probably from having to clean our cesspool several times, but stayed in bed at home.

One day Ellen heard him cry out "Mom!" She ran to see what the problem was, and immediately sensed the presence of the evil one. Together they called on Jesus' Name, prayed for protection under His precious blood, and cast out Satan. Fear fled in Jesus' Name. While we were pioneering in Salta I would at times awake and feel Satanic oppression. Ellen and I would pray in the Spirit, and rebuke Satan. Many of the provincial festivals have pagan origins and are actually paying homage to heathen deities. Satan wants to be worshipped, and hates losing recruits.

Ministry Meditation:

Satan is never pleased when churches are established and people are saved, healed and filled with the Spirit. Missionaries and evangelists are the "lead scouts" of God's "advance troops" to reclaim territories that are largely under demonic control. We engage in spiritual warfare, which Paul describes as "wrestling, not against flesh and blood, but against principalities, against powers, against the rulers of the darkness of this world, against spiritual wickedness in high places" (Ephesians 6:12). This calls for all of God's armour.

Christian supporters who intercede in prayer for those who are "on the front lines", in God's work, are vital to victory. Keep praying! Remember that Satan is a coward, and often preys on people who are sick, like our son Stan with hepatitis, or on people who are suffering mental distress. Surround them with a wall of prayer. And pray for successful Christian workers; they are on Satan's "hit list." Paul commanded "Pray for us." We are eternally thankful for our prayer partners. God is keeping track and you will have your reward along with the missionaries, pastors and others who are labouring for Jesus.

The months passed swiftly, the church was growing, and we had an anointed national pastor. We were pleasantly surprised to have a successful church start in 14 months. But my body was still weak from hepatitis and I missed the triumphal march from the tent to our new church. It was soon time for furlough, but I had a special request to make to our missions director. "Can I stay in Canada for a two-year leave of absence and finish my BA?" The answer was not only "Yes", but also "I think we can keep paying your salary for that time, if you are still available to promote missions among the churches." Wow! Talk about "above what you ask or think" (Ephesians 3:20).

Northern Argentina mineral mountain

Campaign in Jujuy

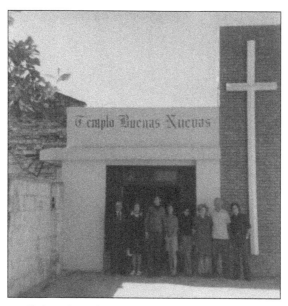

Salta church with national pastor Vena

Home schooling

Chapter 6

University Studies and Ministry in Canada 1974-76

We resumed our studies in SFU, with majors in Psychology and Education. The burden of being a Bible School teacher was still with me, and I wanted to improve my teaching skills. I found the studies stimulating, but there seemed to be very many new theories in both disciplines. A teacher friend bemoaned,"There is always some new idea, that usually falls flat, and we have to go back to the older ways of teaching basics." I also noticed that a large percentage of the psychology students had been seeing a psychiatrist. They were essentially "self-medicating" for their own problems.

One professor gave this existential challenge: "What would you do if someone put a gun to your head and said 'Renounce your faith in God or I'll shoot?'" A nearby student mumbled "I'd say anything he told me to say!" Next class day, I asked the teacher to let me give a response. He assented, and I used the chance to witness about Christ and having firm convictions. The response from hundreds of students was positive, but one Jewish student was offended. The teacher, embarrassed by it all, gave this disclaimer "I myself was a minister, but we don't want such an openly evangelical forum here."

Another professor was discussing social problems, including drug addiction. He was pessimistic about the cure rate, and basically said "Just let drugs cease to be criminalized, and we'll live with that." Again, I asked to give a five minute response, and gave him David Wilkerson's book **The Cross and the Switchblade**. He was skeptical about the high cure rates in a Christ-centered program, but I've visited the original center in Pennsylvania. It was throbbing with life and my heart was moved to hear close to 300 former addicts singing lustily, "I'm a new creation, I'm a brand-new man!"

The Inter-Varsity Fellowship sponsored a debate on Creationism under the auspices of Dr. Gish, a leading Christian apologist. His opponent slunk out before the debate had concluded. Richard Wurmbrand, a Lutheran pastor imprisoned for years in Rumania because of his faith, gave a scathing appraisal of Communism. There are many opportunities to share your faith in the academic arena.

Ministry Meditation

"Higher Education can be dangerous to people's faith, so the Pentecostal Assemblies views it with caution." That was the essence of what Rev. McAlister, our national Superintendent said in an interview with the "Vancouver Sun", in about 1960. This is not altogether untrue. Statistically many youth drift away from the Lord at this time of their life, whether or not they take higher education. In my view, what happens is that people take their "kindergarten faith" to university. They haven't been prepared for the onslaught of unbelief that is almost palpable on some campuses. What can we do to better prepare our youth for the "battle of the mind"? Here are some suggestions.

One, have special youth studies for High School seniors. Polemical books such as "Mere Christianity" (and others by C.S. Lewis), "Evidence that Demands a Verdict" (by former agnostic Josh McDowell) are a good start.

Second, ensure that every student is Spirit-filled prior to leaving for university. My brother John describes a "dark night of the soul" when he prayed and doubts were swept away by a torrent of intercession in tongues. I recall a course in Biology that implied there was no Creator. I prayed and went to the Genesis record to read "and God said 'let the waters bring forth abundantly...Let the earth bring forth the living creature after his kind'" (Genesis 1:19, 24). This was a revelation to me; It is termed "Creative

Evolution." God gave the creative Word, and however long it took, the life process began then. The rest is details for which evolutionists pose a myriad of opposing theories to attempt an adequate explanation. That is simply another kind of "anti-God" faith. As Dr. Gish explained, the main thesis of the second law of thermodynamics is that natural substances almost always deteriorate. They don't naturally grow and evolve into more complex things.

Thirdly, students need to fellowship with other Christians, especially students. Most campuses have a Christian chaplain and Chi Alpha, Campus Crusade, IVCF or other parachurch groups. A nearby church can be a place of refuge. There are now a great number of accredited Christian universities, such as Trinity Western University, near Vancouver, B.C. where several of my relatives have studied or taught. If students keep a strong devotional life, God will preserve them.

Fourth, Christian students need to have a sense of purpose to their lives. For many years the Student Christian Movement was at the forefront in sending out missionaries from annual university conferences. Urbana and other such venues continue with the same purpose. "Give a year and consider giving your life" is a common challenge.

Fifth, I challenge parents to bankroll at least one year of Bible School for their vulnerable children before sending them to a regular university. A firm foundation of faith will withstand a lot of the pressures from hostile university teachers and an unbelieving student life-style. I've personally paid the academic fees for my children's first year of study. They all graduated from Bible School and I'm now starting on my grandchildren. It was William Carey, called "the founder of modern missions," who said "If I had a million dollars to spend, I'd spend it all on Bible Schools." We continue to provide scholarships for students overseas. Over time, I believe this will be our main achievement in winning souls and advancing God's Kingdom.

Chapter 7

Founding Churches and a Bible School in Argentina
1976-80

During these two years at home, we led an almost frenetic pace of life. Ellen took a secretarial course and began working as a temporary employee where needed. We travelled and ministered on weekends throughout the B.C. Lower Mainland and beyond. I figured we were working at least a 70 hour week. Ellen contributed her income to the missions department. At times we'd get a call from the school after a late ferry trip from Vancouver Island. "Your child is very dizzy today, come and get her/him!" With the trimester courses, I was able to study continually and cover many materials quickly. On a very cold day of May 29, 1976 I graduated with honours from Simon Fraser University in Burnaby. Two days later we were on the plane to Argentina.

"Why are you going back?" my uncle Hans and many others asked us. "Haven't you heard of the political upheavals? It's a dangerous place to be." The populist and semi-dictatorial president, Domingo Peron, had died just as we were leaving two years earlier. His wife, Isabella, (a former cabaret dancer, like Eva Peron), had assumed the presidency. Everyone expected the army to take over once again, and it did. We'd seen five different presidents (usually Army Generals), during our first four years, so this was "normal" missionary life for us.

Before leaving Argentina, we had a brief meeting with the denominational "junta" executive leaders. My plan was to return and begin a regional Bible School in Salta. It was almost 2000 kilometers from Buenos Aires, and had the potential of servicing the northern area of the country. On the advice of Paul Brannan, I "got it in writing" that the project had been approved. However, two years had passed, and there was a new slate of leaders. Their mission for us was "Go start some more churches; this time in Mendoza!" Our Salta convert, Carmen, later started that northern school.

Sigh! I had my heart set on getting a Bible School going. To be a missionary teacher was a major goal in my life from my own Bible School days, but, if the national church is truly "indigenous," then it can now also govern missionaries. We yielded to their wishes and set out for Mendoza. This province is nestled beside the Andes, 750 kilometers west of Buenos Aires. The capital city, also called "Mendoza" had a metropolitan population of over a million inhabitants, making it the fifth largest city in Argentina. Spread across a large plain, it is one of the five largest wine-producing areas of the world. (It also has the dubious distinction of having a mental asylum, over 30% filled with men whose minds have collapsed due to alcoholism).

Having grown up in B.C., we loved the Andes with nearby Mt. Aconcagua soaring to nearly 23,000 feet. This is the highest mountain in the southern hemisphere. Across the border our neighbour Chile has many ski resorts on the "snowy side" of the Andes. We did have occasional snow, but winter was mostly a "dry cold".

But where were we going to start a church? We met again a young man we'd known in Buenos Aires. Angel Cruz, now with a family of five, had come from this area. He won a singing contest in Mendoza and had gone to Buenos Aires, the national capital city, to make his fortune. His dreams turned to nightmares as he became a coffee street vendor, with shoes worn out, and the soles lined with cardboard.

But one day he heard the sound of singing and drew closer to investigate. It turned out to be a "street meeting" with gospel songs. They gave him a Bible portion of Mark's gospel which he read all night. Jesus came into his heart and gave him a new song. His fortunes soon turned, but his heart now yearned to see his eight siblings serving Christ. He sold his possessions and with enough money to live on for three months, returned to Mendoza to preach.

The preaching wasn't going too well and the money ran out. Some of his brothers attended house meetings reluctantly. "She's demon-possessed!"

relatives told him when a beautiful but tormented cousin was taken to the meetings. Angel prayed and the evil spirit left, but it kept returning. "Something that holds a link to the devil must be in her home" he reasoned. A thorough search showed up some Roman Catholic cards of saints who are favourites for answering certain prayers, such as finding a girlfriend or boyfriend. Praying to the dead is, essentially, communicating with the dead and forbidden by Scripture. It is no wonder that Spiritism is endemic in Catholic countries. These offending cards were burned and the cousin stayed delivered from demons. Some time later I had the joy of baptizing this cousin, and congratulating her when she got married.

We decided to work with Angel in starting and building a church. His country congregation grew dramatically after the demonic deliverance. Almost the entire family came to Jesus. But there was more good news to come. I felt a special anointing as I preached in the country church one Sunday. Five were filled with the Spirit that day, including Vidal and Luis, two of Angel's brothers. I was thrilled to see the Lord moving in what was a new ministry for us. From that day till the present, Ellen and I continue to see people receive the Holy Spirit. Sometimes over 20 in a single service begin speaking in tongues. Now, as I mentioned earlier, I'll give some teaching on this phenomenon.

Joseph G. Anonby

Ministry Meditation:

As a minister, if you want people to receive that additional power to be witnesses, you need to preach on Jesus' words in Acts 1:8 " But ye shall receive POWER after that the Holy Spirit is come upon you: and you shall be witnesses unto me both in Jerusalem, and in all Judea and and in Samaria, and to the uttermost part of the earth." Jesus further told his disciples to delay obedience to the "Great Commission". He commanded "TARRY ye in the city of Jerusalem, until ye be endued with POWER from on high" (Luke 24:49). Having God's power makes a tremendous difference in winning souls. It was said, during the Pentecostal World Conference I attended in Norway in 2000, that "80% of those who are presently being won for Jesus are won through the ministry of someone who is Spirit-filled." Pentecostals have been called the "Third Force" in world Christianity.

How then, can we preach and seek till we receive and obey Paul's charge "Be FILLED with the Spirit" (Ephesians 5:18)? God does give a special anointing to minister to some, but we can all cooperate with God's Spirit. Here are some suggestions we find helpful for ministering on this theme and seeing people receive the fullness of the Spirit.

1. *PRAY till you build up FAITH in your own heart and the hearts of others.*

2. **TEACH on the subject as an essential for being a better witness.**

3. **PREACH on the subject. People should be reading the books of Acts and Corinthians.**

4. **EXPLAIN what may happen as people begin to sense God's power.**

5. **CALL people to the altar; let there be both seekers and helpers.**

6. **INVITE those who have not received to pass to the front of this group.**

7. **ENCOURAGE helpers, who have received the baptism to assist by laying on hands and praying.**

NOTE: At this point request that all background music cease or be very soft. It is hard to tell if someone is speaking in tongues when music is loud, and difficult to instruct them.

8. **PARDON; God fills clean vessels, so we tell people to request God's forgiveness.**

9. **REQUEST the Spirit ONE TIME (as in "Jesus, fill me with the Spirit"). God does not need constant repetitious requests. Then begin to PRAISE GOD with all your heart.**

10. LAY HANDS especially on those who are visibly moved by the Spirit, or those you discern are close to receiving. Some may be REFILLED and can help minister

11. "SPEAK IN TONGUES" we exhort them. They need to take that step of faith.

12. "LOOSE YOUR TONGUE" we further exhort. They may start with "stammering lips", but they must let God take control of the tongue "the most unruly member." It is not possible to talk two languages at once, so they have to drop their native speech.

13. "KEEP ON, KEEP ON", once they begin the first words of the Spirit. Tell them "Don't doubt, this is the Spirit beginning in 'baby talk'." The words "Mamma and Dadda" are words, though they can seem like stuttering. People who are persistent receive. Don't get impatient. "Some wait longer in a dentist's office to be DRILLED than at the altar to be FILLED!" Jesus said "Blessed are those that hunger and THIRST after righteousness, for they shall be FILLED" (Matthew 5:6).

14. RETURN to them, but have someone else stand in and encourage them to continue.

15. IN CONCLUSION of the altar time:

a. explain again what has happened.

b. ask for a testimony from someone who has agreed to share.

c. tell them to continue speaking in tongues that very night before going to bed. "Just pray till you again speak in tongues. Then do that tomorrow and every day."

d. encourage the pastor to continue to preach and have a place for the Spirit, where people can speak in tongues.

e. tell them that Satan will try to bring doubts, because this is a weapon against him.

f. tell them to seek more gifts, and to USE this gift to WITNESS AND INTERCEDE in prayer. Their "prayer language" vocabulary will increase over time especially while interceding in the Spirit (Romans 8:26, 27).

The provincial capital city of Mendoza is actually composed of four subcities: Mendoza, Godoy Cruz, Guaymallen and Las Heras. Only one of these sub-cities had an Assemblies of God church. We searched the city of Godoy Cruz, the second largest city and found a suitable building on the main plaza. Now, to start a church, as we'd done twice before. But...no church start has exactly the same story. After all the literature and advertising had been done, we still did not have the police permits.

People milled around the building, waiting for the service to start. Pastor Cruz announced in faith "The meetings will begin ten days from now!"

We went again to the police, and gave them "The Cross and the Switchblade" by David Wilkerson. "This is the kind of work we do to help the people" I explained. We had planned to start off with Angel Cruz preaching, but a local preacher/teacher/pharmacist, Lilly Golembioski had another candidate. "I have a cousin, called Jorge Bardey that God is using. Do you want to give him a chance?" We gave him a chance!

Jorge spoke out in words of knowledge, "There is a woman here with an issue of blood. Check it out, God has healed you!" That woman was an elder sister of Pastor Cruz. The miracles and words of knowledge kept coming. A young man told me, "I sense such a powerful feeling when he talks! What is that?" "That's the anointing of the Holy Spirit" I replied. I will discuss this further at a later juncture in this book.

We continued to hold "campaign" services nightly for several months which had become our pattern. As time went on, without stopping the outreach, we would emphasize one night for prayer, another for youth and so on. Gradually the normal church departments began to emerge, as new converts grew in their knowledge and abilities. We also began to build a parsonage for pastor Cruz on the same property.

The Mendoza-area churches invited Yiyi Avila, a prominent Puerto Rico evangelist, to hold meetings in a large stadium. Hundreds came on buses from nearby provinces. But the second night, the police moved in and closed down the services. Hundreds of people began praying in the ditches and surrounding streets. Pastor Angel summoned other pastors and went to the police station to protest. He was informed by a shaken police chief, "Do you think we just do this on our own? We follow orders from above, from the Governor."

In faith, meeting dates were set for a year later with "Cristo Viene" ("Christ is Coming"), Avila's evangelistic association. Huge amounts of advertising literature blanketed the area, but pastors kept wondering "When will permits would be granted?"

As I'm told the story, it went like this. Two days before the campaign, a group of respectful clergy went to the Archbishop of Mendoza and asked for his help in getting a meeting permit. He replied "You just leave it to me. The governor will ask me and I'll say 'It's okay with me', and you won't have any problems." True to his word, we had no problems and hundreds were saved and healed. Dramatic healings occurred. One was sudden weight losses ("lady, grab your skirt!"). Teeth were filled on the spot and I took movies of this miracle.

Angel's brothers, Vidal and Luis, had also been touched by the Lord. Jorge Bardey evangelized with them in two country towns. God moved in a mighty way, and Luis became pastor in an agricultural town called "Nueva California." Vidal lived among the vineyards, near to the county capital town of Santa Rosa. This was a key town but no evangelical church had been able to gain a foothold. Every time a group would attempt to hold a street meeting, the priest would ring the church bells and disrupt their efforts. What could we do to break this spiritual monopoly?

Angel and I hit upon a strategy; we would buy a piece of land and then nobody could kick us out. It seemed like a good idea, till the tall (about 6 1/2 feet) priest heard about it and ordered, "Nobody sell land to those evangelicos!" Finally we found a vacant lot owned by a man desperate for cash. (We learned later that the ownership was questionable, but the deal was done and nobody evicted us.)

But large public meetings outside of buildings were banned, due to the military government's declaration of a "state of siege." There is an Argentine saying "Hecha la ley, hecha la trampa" (make a law, make a loophole). The loophole was clever; surround the public area with a simple single-wire fence. There, you now have a "private outdoor area."

A rickety platform was erected and promo leaflets distributed. The sound system consisted of some huge old "public address horns" that ran on car batteries. You could hear Jorge Bardey preaching in the next county! But when the town cripple danced on the platform that was our best advertisement ever. Then a lady who had a breast removed due to cancer had a new breast given her by the Lord. She went on to nurse two children and later told my wife "How can we not love Jesus? We've begun two new churches." Angel's brother, Vidal stayed to pastor.

In 2017 they celebrated their 40th anniversary, rejoicing in over 30 branch works. His daughter Gladys is helping and asked us to send a voice greeting by internet. The Bible says it right "The fruit of the righteous is a **tree of life and he that winneth souls is wise**" (Proverbs 11:30). We just have no idea how God will multiply miraculously the planted seed of His Word!

Evangelist Bardey never denounced the Catholics. He'd pray "God bless the nuns, the good nuns, the priests, the good priests!" The Santa Rosa priest would drive by slowly, observing these special meetings. Public sentiment was shifting against him. We were told that money had been collected to build a trade school, but he'd spent it on a car for his girlfriend. Later, there was a suspicious death and he was questioned by the police. He shot himself afterwards. Sad, but true.

Ministry Meditation:

How do we preach the gospel in an area that is hostile to traditional evangelical teaching? One very effective way is "power evangelism." Rather than heated argument, this is the equivalent of Elijah's challenge, "the God that answereth by fire, let Him be God" (1 Kings 18:24). "An ounce of experience is worth a ton of logic." I have never found it helpful to denounce other belief systems. People get saved by hearing the message of how to get saved, explained in dramatic testimonies. Anointed preaching "breaks the yoke" and opens hearts. Many people have only a "cultural veneer" of traditional religion which they don't really practise. There is simply no point in offending them; that closes minds. Often they'll begin themselves to talk about hypocrisy and errors they see.

Then, in due time, an earnest seeker may begin to ask sincere questions. At that time I'll point out the scriptures, especially those in Hebrews that speak of a superior, "enduring priesthood" in Jesus. The New Testament law of God's LOVE as He transforms us from the inside is stronger than the external Old Testament law of FEAR and punishment. Catholics have been warned, by the Council of Trent (declarations in 1564, 66), that to profess certainty of salvation is a MORTAL SIN! Hence it is a great relief for them to have a certain salvation, with the witness of the Holy Spirit that they are the sons of God (1 John 5:13). "The WORKS" have already been done by Jesus; "It is FINISHED."

I was still wondering when I'd get into the Bible School teaching ministry. I didn't have long to wait. There was a local night Bible School that needed help. Soon I was travelling to different churches teaching the Bible. We had a VERY primitive stencil that consisted of a small silk screen, stencil paper and an ink-filled roller. One page after another was placed under the stencil. It was archaic, and laborious, but it worked for about 12-15 student at each study centre. Lilly Golembioski returned to the USA, and I became a Bible School Director because nobody else wanted the job.

We now set our sights on starting a fourth church in Las Heras, a southern sector of the capital city. We once again faced the problem of having a prospective pastor lined up. This is how the Lord provided a pastor. Alicia Guzman, our former neighbour's maid who accepted the Lord in Salta, was now completing her final Bible School year. She came to serve with the team that began the church in Godoy Cruz, the northern suburb city. Her boyfriend, Jorge Besso, also graduated and wanted to pastor. Jorge married Alicia, (after we helped him buy a ring), and came to help us in another campaign with evangelist Jorge Bardey.

We only had a rough adobe house for them to live in; the church had not yet been built. We lived ourselves in an adobe (untempered mud brick) house on the outskirts of town. Early one morning we awoke to the swaying of our bed. **"Earthquake!"** I yelled and we all ran out of the house. The dog was barking excitedly and we stood with legs wide apart, like sailors, to keep from falling. I noticed that our simple "fill and dump" swimming pool was sloshing over the sides. While Ellen protested, I ran into the house to grab my "super 8" Kodak camera and film this phenomenon. Our newly-wed pastors hid out in their large closet.

Most earthquakes last under five minutes, though there may be serious aftershocks. We found a large crack in the wall of our bedroom. Adobe houses are the very worst structure for earthquakes. The earthquake epicenter was in the neighbouring province of San Juan, where over 70 died. Shocks were felt as far away as Buenos Aires. I heard there was a sudden large sale of pyjamas.

Our country house had a roof made of thatched cane (similar to sugar cane). A smooth coat of sand, tar and glue made it (sort of) waterproof. We complained constantly to the landlord about the five leaks. Then, in Stan's bedroom, the roof collapsed entirely. By a miracle, Stan was studying in his tree house and was unhurt. After that, the roof was finally repaired, and leaked in only three places.

We kept hearing of the dangers of "mal de chagas"disease. It is caused by a bite of an infected beetle called "the vinchuca." The heart, lungs and other organs are damaged by their bites. Darwin, who travelled in Argentina many years ago, died from "mal de chagas." Where do these bugs love to live? In roofs made of thatch! We found (and saved) two of those bugs from Steve's bedroom. He'd squashed one when he jumped out of bed one morning. We applied the local prevention technique, setting several poisonous candles aflame and leaving the house "sealed" for several days. When we returned, we found an enormous amount of dead insects that fell from the cloth-covered ceiling under that cane roof. We never saw another vinchuca bug after that.

There was one more section of Mendoza city that still needed a strong church. Guaymallen was on the eastern side, and was the fastest-growing area. During a pastoral change, a supervising pastor asked the congregation for a vote of confidence in his leadership. There had been ongoing doubts about some questionable behaviours. Many months had passed in which he never served communion, nor preached. When the vote did not favour him, that dissenting group of people were "put into discipline." "What are votes? They're just pieces of paper!" he scoffed. The district leadership then authorized a new church for these people who had been expelled. We had been able, by God's help to start churches in Godoy Cruz, Las Heras, Santa Rosa and Nueva California. This became the fifth church we'd begun in Mendoza, and now the goal of the national church had three fewer capital cities to reach.

But we now had another national crisis to face. Argentina and Chile have had long-standing border disputes. A famous statue, "Christ of the Andes" stands high in the mountains to commemorate making peace many years ago. Now there was a water-way dispute in the Antarctic Ocean. The southern inter-island Beagle Channel was claimed by Chile as their inland waterway, rather than an international waterway. Also in dispute were the oil rights in waters that surrounded nearby islands. We heard daily military music on the radio, and our windows cracked due to low-level training flights. I was asked if I could serve as a first aid helper, but the official was unimpressed when I replied, "We just pray for the sick!" Finally the Pope, to his credit, solved the problem with quiet diplomacy. Chile got what it wanted.

England also got what it wanted by retaining the disputed "Falkland Islands" (called "Islas Malavinas") by Argentina. These were a Spanish possession, inhabited by Argentine shepherds in the mid-1800's. England, with its vast fleet of merchant and war ships, needed a refueling station in the south Atlantic. They confiscated these islands, which now compose about 70% of the land in Britain's remaining overseas colonies. The new English settlers, under 3000 in all, are subsidized by great amounts of money. England insists, "Only the islanders can vote for union with Argentina." Of course British-subsidized colonists aren't about to vote to cut off their island's income.

In 1982 Argentina fought to reclaim their historic possession. But they misjudged the determination of Margaret Thatcher, "the iron lady." Close to 3000 Argentines died in the conflict. Some froze to death as they camped many nights in the open. Their battleship "Belgrano" was torpedoed and many hundreds of lives were lost. It was an especially treacherous act because the ship was outside the established "battle perimeter" that England had set. Britain's "Harrier Jets", which have wings that can swivel to land like a helicopter were a decisive factor in the battle. American satellite intelligence also played a part, according to Argentine sources. They consoled themselves by saying "We could have beat England, but who can beat America?"

There are two reasons why I believe this humiliating defeat became an unexpected victory in the spiritual realm. Argentines are known for having a huge national ego. They talk of the golden years in the 1920's to 1940's during which they were the fifth richest country in the world in per capita income. England had huge meat-processing plants and imported shiploads of meat. But the Falklands war bankrupted the nation. Inflation soared to dizzying heights. We had 170%/year inflation in our last year, but the rate then jumped to over 1000%. People dumped their grocery carts as stores would announce over the PA, "all prices have now doubled!" Food rioters ransacked the supermarkets where we shopped. But with humility and poverty, proud Argentina was ready for revival. I will take up that story in a later chapter.

The indigenous church is defined as "A church that is self-governing, self-financing and self-propagating." Melvin Hodges, an Assemblies of God missionary to Central America, wrote **The Indigenous Church** to describe the sometimes painful process. As a church passes from infancy to "adulthood," the founding missionary needs to cede his place, as I've written earlier. Leaving an entire denomination "on their own" is a much larger step. This is what the Pentecostal Assemblies of Canada decided to do in Argentina, the West Indies, and Liberia at this time. The idea was to enter "unreached people groups," especially in the "10-40 window" of hard-to-reach countries. Many of these are Islamic in their beliefs.

The idea was that our Canadian missionaries would be re-deployed elsewhere. Years before, we had been advised by West Indies missionary Harold Eggleton, "When your children get older, go home and stay home till they are settled. Then return to the field." We sold all our possessions, were feted and said our "Adioses." But about two years before, while in my daily Bible reading, a verse just leaped out at me. God spoke to Jacob as he left to journey, "And behold I am with thee and will keep thee in all places whither thou goest, and will **bring thee again into this land**; for I will not leave thee until I have done that which I have spoken to thee of" (Genesis 28:15). A year later, I happened upon the passage again, and felt the same surge of joy in my heart. As our plane soared over Mendoza I turned to Ellen and through our tears said, "We will look back and say 'these were golden years'."

Chapter 8

The Daily Life of a Missionary Family

There are some things that every Christian family has in common. We all work, eat, sleep, study, worship and play. However, depending on where a missionary serves and the kind of work they are involved in, there are often major differences. I will delineate these in the above order as we experienced them.

Take the matter of **WORK**. Ministers are often told, in conferences and elsewhere of the "proper order." "God first, then family, then church." On first glance is seems like sage advice. But I beg to differ. Yes, God is undoubtedly to be first. But I see a danger in putting the church in last place. Christ is the head, and the church is His body. Do children get priority over Christ's body? For me, this is a false dichotomy.

Instead, this is what I propose: put church and family together. Establishing churches and winning souls was the call of our entire family. Our children would sing in street meetings and who could resist our "little blondy" Joy, at five years of age handing out invitations or tracts? If they dropped a tract, she 'd pick it up and run after them saying, "Look, you dropped this!" Our children made friends with neighbours, won them to the Lord, and invited them to church. At age ten, Steve taught a Sunday School class in our car. Stan would help me hand out chorus sheets and run the movie projector. They got involved in the church construction projects. We were not always able to "protect them" from observing demonic manifestations. And they were with us as people were led to salvation, healed and filled with the Spirit. Church-founding campaigns would go on nightly for many months and our children were usually there.

Fast forward a number of years, and each of them went to Bible School and entered the ministry. Steve had some dramatic experiences with demonic deliverances. We had them helping us pray for people to receive the baptism, and soon they were doing that on their own. They all know how to lead a soul to Christ, and they continue to minister.

Ministry Meditation:

Do YOU KNOW HOW TO LEAD SOMEONE TO JESUS? Many people are not sure of their own salvation. I can recall my own uncertainty as a child. I'd had a dramatic vision or dream while very young. So this young "Joseph" told his brothers, "I dreamed and saw a beautiful rainbow, and written on it was the message 'Jesus is come'!" My brothers weren't sure that was the proper grammar, but I recall the vision to this day. Still, there came moments of doubt, because Satan is "the accuser of the brethren." One moment I'd be singing lustily "When the roll is called up yonder I'll be there" while picking huckleberries. Then I'd feel unworthy and kneel by the bush to pray (again and again) for Christ to save me.

It was my own brother John who helped me gain assurance of salvation. It happened during the two week campaign in which all our family received the baptism in the Spirit. John explained simply, "But as many as RECEIVED HIM, to them gave he power to become THE SONS OF GOD, even to them that BELIEVE on His name" (John 1:12). I had received and believed, it was as plain and rational as that. A few days later I received with my powerful baptism an inner witness of the Holy Spirit.

Years later, I'd see people in Argentina coming to the altar for salvation, yet speaking in tongues as they made their way down the aisle. They were saved, but had no assurance. I wrote a tract with this simple "logic and assurance of salvation" which many found very helpful. Parts of it have been termed "The Romans Road."

1. *Romans 3:23 "ALL HAVE SINNED and come short of the glory of God." That means me, and everyone you know has*

sinned. Normally the person to whom you are witnessing will admit that they, too, have sinned.

2. *Romans 6:23 "The WAGES OF SIN IS DEATH, but the GIFT of God is eternal life through Jesus Christ, our Lord." Our work for sin brings pay; it is the penalty of eternal death, hell and separation from God. But God has a GIFT; and it is not something we deserve or earn by our good works. If good works were sufficient, Jesus would not have to die on the cross.*

3. *Romans 5:8 "But God commendeth His love towards us, in that while we were yet sinners, CHRIST DIED FOR US." In other words, the debt and death we'd "earned" by being sinners was paid by Christ, the perfect sacrifice in our place.*

4. *I John 1:9 "If we CONFESS our sins, He is faithful and just to forgive us our sins and to CLEANSE US FROM ALL UNRIGHTEOUSNESS."*

I then have them pray after me a simple prayer: "Jesus, I confess I'm a sinner. Thank you for dying on the cross to take away my sins. Right now, Jesus, forgive ME, take away MY SINS and come into my heart. I want to serve YOU from now on!"

Now is a crucial time to instill ASSURANCE of salvation. Ask "Does God lie? Is the Bible true?" Then say, "God made a holy contract here. He said 'confess your sins and I'll forgive your sins.' Did you confess your sins? And what does the Bible say God did? He forgave your sin. Welcome to the family of God!

I normally add this counsel, "You have planted a seed, now this is what you need to do to make that plant GROW and bear fruit, so you'll be a GREAT CHRISTIAN." Just follow the steps of Jesus.

1. *Jesus prayed every day.*

2. *Jesus read the Bible.*

3. *Jesus went to church, (even though it was far from perfect).*

4. *Jesus testified to His faith.*

I eventually wrote a (Spanish) tract for prospective or new converts to help them get established in the faith. Nobody need live in fear that God didn't hear their prayers.

Returning to our theme of **WORK** in a pastoral or missionary family, my advice is to make your children your "apprentices" in ministry. Today the term most often used is "mentoring." In the Old Testament, the priesthood was an hereditary office. Missions work is not something that "Dad and Mom do." It is what the whole family is doing! Stan, at age 14, reportedly told a visiting group, "I'm a missionary here in Argentina." This is not to say we force our children into ministry and make them feel second class if they don't follow our calling. But we provide opportunity for them to be involved with us in eternal work, and they'll never forget it. "I think going to church is a blast" Steve would tell his Canadian school friends. Again, I affirm my belief that there should be no priority of "children before church." Combine the two.

Another aspect of work is the daily "housework." This often has substantial differences from how things are done in North America. As one missionary pointed out,"At home, we have a lot of "mechanical servants." (Such as vacuum cleaners, dishwashers, specialized ovens, automatic washing machines, running water, phones, fax machines, internet, and many other things we take for granted). At the time we lived in Argentina, we had NONE of these things on a reliable basis. Dishwashers were simply not available, though that would have been a great help when we hosted large groups. Our ovens were too primitive to precook food and during winter in Buenos Aires, gas pipeline flow was so minimal that even the stove gas pilot lights went out. Automatic washing machines would usually not work because of an unreliable water supply. Electrical voltages would often drop,till our lights looked like candles. Then they could suddenly spike and ruin the motors on our appliances. (But we at least did not have a servant beating clothes on a rock in the river, like earlier African missionaries).

The water was pumped up from 150 feet deep. Septic tanks were 40 feet deep. You get the picture; we had to boil all our water. The large 5 gallon water jugs one sees today were not available in 1970. And if you drank the water, you almost always got parasites. Parasites caused digestive problems, and gave you fetid breath. Several missionary colleagues had babies who did not survive due to parasitic infections. We drank large amounts of Coca Cola, but, as Grandma Bonderoff predicted, it ruined the teeth of Joy, our youngest child. In Mendoza we had a deep well in the country so that water was "safe." However, the pump that drove the water into the vat atop our house kept breaking down.

Phones were not very common. In Mendoza, Argentina it cost me $800 (about two month's wages), to install a phone in my name. It cost another $800 to return the phone to the owner's name. And so few people had phones in the country that their usefulness was marginal. In Jujuy we used a "public phone" beside the police station. One would ask the attendant, "What is the wait time for a call to Canada?" It was usually a couple of hours before a "window" of time emerged. All these "mechanical servants" save people large amounts of time.

The solution, on some mission fields, is to employ some household help. On some older fields, a "houseboy" would do the buying, cleaning and cooking. Another might be assigned to tend the garden while maintaining and guarding the house. They usually lived behind the missionary house, and considered this their "lifetime job." A new missionary got "a house with servant." It was unthinkable to fire such a servant. Many became Christians and would literally give their lives for the missionary. In return, the missionaries were able to give themselves to winning souls and training workers. It costs the missionaries additional personal finances, but winning souls is why they became missionaries. I hope this helps people understand what having a household servant implies on the mission field.

Next we look at how we **EAT** on the mission field. When we first arrived in Buenos Aires, there were very few supermarkets. Ellen would go with her dictionary and buy milk at one shop, meat in another, vegetables in another, and so on. It was usually done daily, and took a lot of time. She pulled a wheeled basket home behind her, like most shoppers would do. There were few "prepared foods" so getting a meal together was a considerable task. There were only two "dry cereals" available: rice crispies and cornflakes. The curious check-out cashier asked us, "What do you do with these flakes?" On the box of cornflakes was the suggestion "use it with milk or ORANGE CRUSH!"

Argentines were not accustomed to freezing meat; it was bought fresh, daily. Meat is a large export/money earner for Argentina. In the early 70's, local consumption was so large (a pound a day per person), that the government imposed rationing. People panicked because they were only allowed to buy meat every other week. They were astonished that we would pack our fridge's freezer compartment with a week's supply of meat! "Won't it go bad or taste terrible?" The Argentine "asado" (barbeque), is the expected fare at any picnic or family gathering. While the men gather around the grill, varied pieces keep being put on. Some are from visceral organs that we would not consider eating in America. We ate a whitish stew that turned out to be cow's brains once! It is not a dish I recommend.

Especially in northern Argentina, they loved to eat "locro." This was a cow's udder, mixed with corn and vegetables in a thick soup. The udder was often hard to chew, but it provided a lot of flavour. On national holidays it is common to see a load of wood or branches atop a car, along with a home-made steel grill. The family would light a fire anywhere on the ground and cook on the little grill above it. If you happen to pass by someone eating it is **almost compulsory to offer food**. A common rhymed saying often repeated is, "El que coma y no convida tiene sapo en la barriga." This means "He who eats and doesn't share, has a toad in his tummy!"

The Argentine eats a breakfast of bread and jam with coffee or hot chocolate. The children also drink coffee. When I once protested, "But coffee will stunt their growth!" the parents looked at me. They looked at their daughter (taller than my boy of equal age), then looked again at me with puzzlement. "Okay, now who believes in 'old wives' tales?"

Their noon meal is usually between one and two in the afternoon. It is a **very large** meal, in which children are served portions far larger than

I can eat. When one is a special guest, the meal can go on for nearly two hours. Lots of conversation and comradery. Argentines have a keen sense of humour, and **never groan at a joke**. It is polite to say "that was a good joke" and keep everyone talking.

About 5 in the afternoon is the time for drinking "mate", a tea-like herb put in a gourd and sucked through a common steel straw with a sieve at the bottom. Sugar is added after each person drinks, then passes the gourd to the next in the circle, **without washing the straw**. If you wipe off the straw it is considered an offense to the previous drinker. Only in Mendoza did we see some people pouring hot water over the straw before passing it on to the next person. "You learn to speak Spanish by drinking mate" is their saying. We learned the common speech and expressions in this casual context. Sometimes I'd try out an expression I'd heard on the street during a sermon. I'd then see a deacon look up, wide-eyed, before writing it down. He'd tell me afterwards, "That's 'street language', best not used in church!" We were happy that our children started speaking Spanish with their playmates. Until church visitors asked, horrified, "Do you know what your children are saying?" "Ah, no... sorry about that!"

"Mom, we're in trouble!" Steve exclaimed. We had a "welcome meal" for a new pastor at our house, and the main menu was "empanadas," a sort of meat pie. To our surprise, almost all the church showed up (far more than Ellen had expected). Yet, to our astonishment, **the empanadas just never ran out**. We have no logical explanation. But God had supplied boxes of groceries "just in time" many times. He supplied us with deer meat for the winter. The "empanada miracle" was more dramatic, but God supplies by many means.

The children in Argentina go to bed very late. Then the adults eat the last meal of the day about 10 pm (or even later in Spain). This is a "real cooked meal" not just sandwiches. One missionary, Mrs. Trussel, was hosting a group of Argentine youth during a campaign. Later she found her evening sandwiches in the garbage can. They were offended, but the Argentines were also offended to think that they'd not been served a "proper supper." Some missionaries continue to eat on their usual North American schedule. They invite people over for supper at the time most Spanish people drink their mate or have a coffee. Nationals don't eat two large meals one after another, and are frustrated. We recommend "going native" with eating hours; it's no big deal to change again when we return home.

Sleep is a third element of daily life. And who has not heard of the Spanish "**siesta**"? In Argentina they have a saying "La siesta es sagrada" ("the siesta is sacred"). Since people often go to bed after midnight, and the work day begins at 8 in the morning, tiredness sets in after the big noon meal. The hours of about one to five, depending on the season, are for siesta time. People put on their pyjamas and go to bed. Stores are closed, street lights go on "blink" in interior towns, and children are not allowed to play on the streets. At five, it is common to see people, still in their pyjamas, having their mate on the sidewalk in front of their house. Homeowners pay for their own section of the sidewalk, so it belongs to them.

It was useless and very impolite to phone or call on people during the siesta time. We adapted by enjoying our family meal, followed by an extensive devotional time with our children. Then we'd all take our siestas for maybe an hour. Afterwards, I'd have my personal devotions and study time. The years have passed and we still go to bed after midnight. And I still like my (much briefer) 20-25 minute siestas. I find it relieves the tensions of the day, and "The Woman's World" magazine recommends it too!

The fourth element of many families' life is **Study**. In Argentina many children begin nursery school well before kindergarten age. Stan went to grade one in a Spanish school where our landlord's wife taught. This school had a morning and afternoon shift, as do many schools. Children all wear a white overall uniform, because Sarmiento, a historic President and educational pioneer, wanted to foster equality. School was free (sort of), but nearly every day we were told we had to buy something else for art, or some other project. There were even specialized stores that sold different kinds of paper. Moms routinely do the children's extensive homework. Ellen was also teaching Stan the British Columbian Provincial correspondence courses at home. Finally she got to her limit of patience when she was asked to buy "green monkey fur" for a project. We yanked Stan out of his first grade and took turns teaching correspondence.

This solved the problem of a school year that was six months "out of sync" with the North America system, since Argentina's "Summer" is the northern hemisphere's "Winter." Some missionary children lost two or more years of school during various furloughs. Many missionaries have grieved, along with their children as they were sent to distant boarding schools. This was a standard policy with some mission agencies. In my opinion this was **never necessary** but has often been pictured as a needed sacrifice that a truly dedicated missionary should be able to make. Missionaries and their children should be entitled to make their choice. Children usually adapt, but some have been traumatized and are bitter towards their parents and missions in general. There are some very obvious alternatives.

One is teaching a child by "home-schooling." Usually a recognized correspondence course is used, but parents can buy books and do it themselves. In Salta the missionaries formed a kind of school co-op, that taught both missionary and other expatriates' children. If missionaries feel very inadequate, there is often a bilingual national who will be happy to do the job. The parent is then freed to do other missionary work. Sometimes a tutor comes from America with the missionary task of teaching one family. We hired several national bilingual teachers who turned out to be poor tutors. They couldn't resist doing the children's homework, and leaving work early. Finally Ellen taught the children and hired a maid with the "teacher wages."

We had, however, a problem with Joy and her reading skills. Ellen was praying for God's help. Meanwhile in B.C., Alice Leland, a school teacher and former pastor, felt God saying "Anonbys, Argentina, Anonbys, Argentina!" She consulted with her pastor, Marvin Forseth, and he told her "Go!" She was a reading specialist, trained to help dyslexics by precise techniques. In short order, Joy began to love reading and devoured several religious-themed books. Alice gave up her school salary and paid her own way to help us for an entire year. She is one of the quiet heroes that are lifetime partners in winning lost souls. It was not easy for her suddenly to lose her pupils when a child would yell "the chickens are loose again!" She sometimes would sigh, "I'm no spring chick!", but she stayed on the job and became a life-long friend.

Worship is a vital part of every Christian's life. At times, a missionary family has both unique challenges and singular blessings. Family devotions are, I believe, a basic building block to pass on our faith to the next generation. This was how I at the age of six was drawn to accept the Lord as my personal Saviour under my Father's guidance. Our children made their first prayers repeating bedtime and dinner prayers. We'd sing age-appropriate choruses, and use an illustrated Bible in our family devotions. One Bible came in comic-book form, and that was a great hit. They kept wanting to read more chapters to see what happened next. I'd play my guitar as we sang.

This brought an additional benefit. One by one, our children began to harmonize. Joy recalls how I'd sing alto into her ear. Our children took both singing and instrumental music lessons. Ellen and I sang duets for years, especially while we travelled in missions promotions. Later, our children joined with us, till we had a four-part harmony going. We had often been asked "Can you sing?" Then, "Can the family sing?" But now the question was "Can your children sing?" People would cheer at their enthusiasm and melodious harmonies. As the years have passed by, each of the children is involved in some kind of music ministry. The Lord has given me many missionary songs, which we now sing as we travel (once again just the two of us), in missions promotion.

We also had the children memorize scripture. "Thy word have I hid in my heart that I might not sin against thee" (Psalm 119:11). After a Bible reading, we'd ask questions to see what they understood. They were surprisingly perceptive in applying the Word to everyday life. Once the Word has been internalized in a child's heart, it's a lot easier to be a parent. You don't have to be constantly saying "No!" when the mind has been programmed by the teachings of the Bible. The Bible contains "God's rules" for the entire family. We'd go to both "regular" church and to weeks-long evangelistic services as a family, with few exceptions. "Baby-sitters" are not very common overseas. Our children would join in all the activities of the church, helping us whenever possible. Stan started a choir, and played his guitar. We were not usually able to "shelter" them from demonic manifestations, but would discuss these things later.

We wanted our children to sense the presence of the Lord and the moving of His Spirit. I'm not in favour of "Children's Church" always displacing the entire "church worship" experience. In one such service, Pastor Forseth called out "Anyone else want to be baptized?" Steve, at age eight, marched to the front to be baptized. Years later Pastor Jamieson, in Richmond, gave a similar challenge, and our ten-year-old grandchild, Sophia, marched forward to be baptized. Children are often sensitive to the Spirit.

"Here's a 'Pentecostal handshake' Mrs. Forseth" said Steve, about age 8, one Sunday morning. Our pastor's wife, Dorthy Forseth was arguing with God about His "nudge" to talk to someone about a spiritual struggle they were having. Finally she put out this fleece, "If someone gives me a 'Pentecostal handshake' of 25 cents, I'll obey You." Guess what coin Steve pressed into her hand, moments later? My brother Daniel talked about our children "living in an enriched spiritual atmosphere." Yes, like the child Samuel, children are open to God. But parents can help set the stage by exposing them to worship and demonstrating love for God in their personal lives.

It's important to make allowances for different styles of worship. Many of our Spanish friends love to have **very loud music**. Everything is loud and boisterous, including their gestures. As one gets older, these loud noises are harder on your ears. I simply tear off small pieces of kleenex and stuff them into my ears. I'm not going to change an entire society. Street music at Dominican food/drink stores blares out most every night. It is supposed to stop at midnight. On weekends it is allowed later yet. Meetings are often **very long,** about four hours or more on their monthly "jubilee gatherings" with branch churches. You can come later and this is acceptable. Many people take turns in leading a couple of choruses.

A strong contrast appears in our "Methodist Pentecostal" services in Chile. When one is feeling the presence of God, you call out "Let's give glory to God." And the people rise up and repeat three times "Glory to God, Glory to God, Glory to God." In many fast-growing Korean churches, people worship until the pastor sounds a little bell on the pulpit. Worship stops at once. We expect to adapt to new cultures on the mission field. Some of our American churches are dividing over issues like these. If I had to leave a church due to noise, I'd have nowhere to go in the Dominican Republic. Learn to adapt to differing church "cultures" and be sweet!

Ministry Meditation:

Some people are upset that church worship is "not the same" as it used to be. They dislike a youth element setting the pace of music. Some choruses are disdainfully called "the 7-11" = "same chorus with seven words, repeated eleven times." But looking at history, we see the church from Bible times, "singing a psalm." The "Psalter" was the songbook of the Protestant English churches. A young man, complained about the strained wording and music. He was challenged "Go and do something better!" Isaac Watts accepted the challenge and changed the face of church music for two centuries.

But in the interest of equality and fairness a church should not let youth disenfranchise their elders who paid for the church in which all are worshipping. I have observed some churches that bridge that gap by having both senior and youthful people serving together as worship leaders. A wise pastor looks for a mixture of old and newer songs. When we sing a traditional special song, we note how older folk really begin to worship God, as they recall good memories of times past. Some churches have a "traditional" service, followed by a "contemporary one. In general, when people maintain a spirit of love and obey the golden rule, peace prevails.

Play is the final area of missionary life I want to touch on. This can be a great "family time" to build fond and exciting memories. We had a "normal schedule" of home-schooling from Tuesday to Saturday. Sunday was church-related activities all day. But Monday was "family day off." Even our dog would try to squeeze into the car on Mondays (but not on Sundays)!

About our dog, that's another story. Joy had been pleading for a dog for some time, but we lived in the city. We had a tiled back yard and our front door opened right onto the cement sidewalk. It didn't provide any place for "a doggy to go." Joy began praying for grass to grow on the sidewalk. One night, after a meeting in the country, she came to us with shining eyes. "God answered my prayers, He gave me this little puppy!" Who can break a child's heart and her trust in God?

We called the (female) dog "Guina." In the Paraguayan Guarani language, this means "girl." Our landlord was not happy with our dog. (Most of our Argentine landlords were unhappy people). People would say, "If they are rich, they are bad." I almost came to agree with their assessment, but that's another story. We moved to a rustic (adobe) home among the vineyards in Mendoza's outskirts. There our dog had lots of space to roam, and Guina became our children's most trusted friend. We learned that people were saying we had a fierce dog. Not true, but why contradict them? She once barked us awake to give the alert of a thief. He fled with a few pesos from a purse left in the car. In due time Guina populated the neighbourhood with her frequent puppy litters.

On our days off we usually took different roads to places we'd not seen. Our favourite places were by mountain streams where we'd eat a picnic lunch or roast some weiners. Sometimes we'd take a book to read out loud. The children would make dams and splash in the streams. We also would climb the hills and even found a place to fish for brown trout. We had to keep an eye out for rain storms in the high Andes. The lower-level highways would dip into "dry gulches" with a measuring stick at the lowest point. Very suddenly, a torrent of water from rains in the distant mountains could come down that gulch, flooding the road and making it impassable for several hours. Scores of cars would wait for water levels to abate before daring to cross. Every year people would drown by trying to cross when it was not safe.

Have you **ever crossed a train trestle bridge driving your car**? I didn't think so, and I don't advise it. But on one of our outings we drove up a rough small road over a pass in the Andes. The rudimentary map showed it joining up with a main highway after crossing the Mendoza river. On the mountain pass we met a group of youth driving a jeep through the foggy heights. The larger cacti had remnants of hoar frost running off their trunks. The youth stopped us and warned "Go no farther, the road is bad!" Being raised among the mountains of British Columbia, that was a challenge for me.

Our food was running low. The gas was running low. And we got a flat tire as we drove down the road that was basically a dry river bed. The road had a poorly-marked division, and we took "the road less travelled" as the poet said. And indeed "that made all the difference!" The road **ended at the river and the train trestle**.

Now, what to do? We were almost out of gas, so a return trip was out of the question. Could we dare cross the bridge? What if a train came? Looking closer we saw that people had already tried that, so the railroad had attached huge spikes sticking upwards on entry plates. But someone else had removed some spikes, so these protective plates could be swung aside. We took a deep breath...and drove the car up on the train bed, straddling the narrow gauge tracks with the help of my car jack. A group of tourists on a bus across the river came to our aid. With Ellen praying out loud, (and filming the crossing on our 8 mm movie camera), we bounced across the trestle and onto the highway on the other side! I don't think I'll recommend making that trip again, but once was fantastic!

We let our children have a fair amount of liberty to seek adventure on their own. Stan and Steve took their bicycles high into the Andes mountains. Huge condors with 12 foot wingspans would land close to them. These birds would sometimes fly off with kids (of goats), and the goat-herders hated them. Stan yelled "Come Steve, I've got this big goat cornered, do you want to ride him?" Steve took one look at the billy-goat and decided against the challenge. They'd stay overnight with goat-herders and come back from their adventures smelling like goats. But oh, such glowing eyes as they recounted adventures that none of their childhood friends would ever have.

A final "extra" of missionary "play" we enjoyed was travel to exotic places. Before Alice Leland, our children's teacher left, we took her for a celebratory holiday to "Barriloche" in southern Argentina. ("To Barriloche and/or BUST"). We pulled a homemade trailer that kept breaking down. A gasket blew when we towed the trailer over the Andes and several tires were shredded. Ellen read a book by Joni Erikson, a paraplegic who got saved, to pass the time and calm the children. Barriloche is close to the glacier district of Patagonia. It has lots of Swiss immigrants who make delicious chocolates. Plus many Welsh folk who tend flocks of millions of sheep.

For very little additional personal expense, once we were en route to Argentina, Spain, or home to Canada we would stop off at unusual destinations for a day or two. That's how we had some unique "mini-holidays." We took a train that navigated "switch-backs" (alternating forward and reverse) over a mountain to visit the famed "Machu Pichu" in Peru. We climbed the equatorial highlands of Ecuador and the "sun and moon" pyramids of Mexico. We toured Israel once for less than it cost to return on a direct flight to the Canary Islands. Our children enjoyed trips to Disneyland CA and Disneyworld FL plus the beaches of Rio de Janeiro, Brazil–all on journeys to and from the mission field. Sometimes we'd preach and sing in these exotic places, and this added a new dimension. In Trinidad, we concluded that the congregational singing was so marvellous that it far surpassed our "family special!" These are some of the **joys of a missionary family at play!**

Chapter 9

Resettling in Canada (1980-84)

We were now in the "re-entry phase" once again. You may experience "culture shock in reverse" when you return home after four years away, with few visitors from "home." My dad and Mom had visited once, while we lived in Jujuy. Dad was admiring a llama when it spit at his face. He tried to flush a toilet, and got a shower instead. We visited some Norwegian missionaries and I found myself translating into Spanish, English and Norwegian all at once. Not always too successfully!

We stopped en route home to see the wonders of ancient "Machu Pichu" in Peru's Andes. An ancient fortress city in the high Andes, it was never found by the Spanish conquerors. In Los Angeles we stopped to see a Bible School friend, Bob Wilson. While we waited for him to finish his day at his Christian book store, we took a motel. There were deep shag carpets and a colour TV. We were over-whelmed. After living under a leaking asbestos-cement roof for years, and shivering inside the house, this seemed like heaven. I was overwhelmed.

I went aside and fell on my knees. What had I been missing all these years? The Lord spoke to me powerfully with two scriptures. **"Love not the world, neither the things that are in the world"** (1 John 2:15). The other verse was "**If ye then be risen with Christ, seek those things which are above...not things on the earth**" (Colossians 3:1,2). It is very easy to become accustomed to what we do not recognize as a luxurious lifestyle, until we undergo "culture shock" in overseas primitive conditions. But less understood are the "re-entry culture shocks" of seeing food wasted, clothes discarded, cars scrapped and furniture junked. We see North Americans missing opportunities to invest in winning eternal souls by not living a simpler life. In a later chapter I'll outline a workable plan that can dramatically increase your missions giving as one way of obeying the great commission.

We moved back to Vancouver B.C., where the Lord had helped us buy a house some years before. Four tenants, each renting a small suite, had been paying the mortgage. We'd built a seven foot square storage unit in the basement to hold our things during our last four years in Argentina. I still felt God wanted me to be better prepared for Bible School teaching. So I took what I expected to be a five year leave of absence to upgrade my education and get our children to re-adapt to Canadian culture.

My plan was to attain a Masters degree in Psychology. I was told "only about 3% of applicants are accepted. But a masters in Educational Psychology accepts about 5% in the Regina University." I applied to the head of the psychology department, who had been a student under my brother John's teaching at UBC. A local Regina church, which had undergone a division, was looking for a pastor. Their first question was "What do you think is the role of the church board?" I think they wanted me to answer "the board is in charge of the church," not "The board works with the pastor." No wonder they had such a history of pastoral desertions and church splits!

Meanwhile, we waited and waited for a notice from the Psychology Department. Nothing. Then a call came from Central Bible College, "We need a missions teacher, would you be interested?" On the basis of only this tiny door of teaching one subject, we said "Yes" and packed to go. We couldn't believe how much "stuff" we'd collected in just one year. We packed full a 23 foot moving van and Ellen drove the car to Saskatoon, Saskatchewan.

God had already provided a place to live. This was how it happened. When we sold all our belongings in Argentina, inflation was so high that we got more for used furniture than what we'd paid for it new. We "parked" this money for awhile and waited for God's direction. While itinerating in Saskatchewan, I met a country pastor, Rev. Shultz. He'd lived and ministered in Dundurn, 20 miles south of Saskatoon, but was now pastoring in Cutknife.

His father-in-law had renovated a small house in Dundurn to make three bedrooms. They were now ministering far away, couldn't sell the house, and were too poor to buy car insurance. I felt led to pay for their insurance, and they gave me a picture of the house. When I moved to Saskatchewan, they sold me the house for a modest price, and took back a low-priced second mortgage in an inflationary year when rates were over 16%. God provides; "give and it shall be given unto you."

But how were we going to pay living expenses with three teenagers to support? And how many years would it take for me to get a graduate degree? The University of Saskatchewan was very pessimistic about chances for entrance into the Masters in Psychology program. However, Central Bible College had an affiliation with the Lutheran Seminary, and were willing to recommend me as a student based on my previous GPA. They wanted their teachers to have studies beyond a B.A. I applied for some cleaning contracts and was promptly hired by a subcontractor. This was the start of a series of miracles.

I applied and was accepted at the Lutheran Theological Seminary (LTS), just days before classes began. **Two days later** we got a letter that had followed us through several address changes. **"You have been accepted into the Master of Psychology program at the University of Regina!"** Too late, too late! I was already enrolled in the M. Div.,(Masters of Divinity), program. God leads by both open doors and shut doors.

Many people tell me "I'd like to go to Bible School and enter the ministry, but I have a family to support." My response is, "I had a family to support also, but if God is in it, He can supply for you too!"

Here is a partial list of some of the financial miracles God did for us. 1. The LTS agreed to pay me a $300 per month interest-free loan for my living costs. 2. They gave me free tuition. 3. I told the unemployment office that I needed to upgrade my training to get a new job. The officer instructed me to fill out "studying" on their form, and I'd be paid unemployment. 4. The Bible School paid me monthly for teaching part time. 5. A nearby student drove me the 20 miles to town daily, for a modest gas donation. 6. The monthly mortgage was less than an average rent. 7. We got many week-end ministry engagements in which churches paid us honorariums. 8. Some anonymous donor gave us several substantial donations. 9. Ellen became a very successful "Tupperware" salesperson. 10. When our car was crashed into and totalled, the insurance company gave us a good settlement, because I showed them all the (many) repair bills. 11. A pastor in a church where we had preached, heard about the crash and gave me his car instead of trading it in for a newer one. Isn't God good!

The only stress point was mastering Greek studies. I found myself spending more time on Greek than all the other subjects combined. Something had to give. I gave up my cleaning contracts first of all. Then Miss Brown, a Greek teacher at Central, offered to tutor me for free. I temporarily dropped Greek and went full bore into the M.

Div. Studies. They were completed in record time, except for the Greek. A common class saying was "I wish the Greeks had toothaches for weeks."

There were five Pentecostal students that were studying at LTS. The seminary was happy to have us, as their attendance was low. Some of the professors were fairly liberal in their views. After one such class a Lutheran student who had been Spirit-filled talked to us at lunch hour. "I was a physical and emotional mess, with my faith about to falter due to spiritual doubts. Then I prayed about it, and decided 'This is garbage' and dumped my doubts. Peace and health returned." Dr. Birch, Central Bible College director would tutor us and quote new Christian conservative scholars. Many of the textbooks quoted German theologians who assumed miracles didn't happen. They "late-dated" prophecies. This removed the supernatural element. But the recent findings of the "Dead Sea Scrolls" and more rigorous studies disproved these "late-date" theories. And, given my experience with the supernatural on the mission field, I wasn't about to trade in my faith for theories.

In near-record time, most of my formal studies were completed. (I completed the Greek requirement later during a "church Christmas week break"). Unemployment insurance was running out. I needed to get a job. This transition once again to "no income" could have been a nerve-racking time with three hungry teenagers to feed. As I prayed I felt the Lord telling me, "I will supply if you worry about it, and I'll supply if you **don't worry about it!**" That was an easy decision to make!

Anonby Adventures in Missions

I was willing to take even a small farming-town church in order to continue teaching. One church at which I was a candidate was exactly that: small town, small church, smaller parsonage and small salary. While I awaited their decision, Superintendent Roset, from neighbouring Alberta, told me of two churches that needed pastors. Both were eager to hire me.

On the plane bound for the second church, north of Edmonton, I sat beside Demos Shakarian, leader of the Full Gospel Businessmen's Association. It was a joy to hear this godly man talk about the revival of the Spirit in many areas. He also asked questions about our missions experience in Argentina.

We decided to accept the call to a northern Alberta church. It looked like a fine town of about 4000, including a large outlying farming area that stretched up to 30 miles in each direction. A nice three bedroom parsonage with a finished basement stood across the road from the church. About this time a deacon from the tiny Saskatchewan church phoned to say, "You've been turned down. I'm ashamed to say we've turned down 23 candidates!" He was even more dismayed to learn I had been accepted at a church with at least four times the size of their congregation–plus a house and salary that were both twice as large. I had been willing to pastor the tiny church, but God had something better.

We put the Dundurn house up for sale, but were asked if we could rent it instead. Stan now had his driver's licence, so we drove a moving truck and two vehicles to our new pastorate. We were met by two members of the church board who had children about the same age as ours. In short order we were moved in and the children settled into school.

From the very first Sunday, the Lord began to give me very strongly-anointed messages about committed Christian living. One person, who later admitted to infidelity, said words to the effect of, "Either things will start to change around here, or you'll be voted out!" Country churches have a different dynamic than city churches. Farmers don't move every five years, like the average North American. There is more intermarriage among the congregants. They fight a lot, but they also stick up for "family," despite some doubtful behaviours. They'd sooner change pastors than make changes in personal behaviour or the church political power structure. But all this took awhile to figure out.

I thought that my dream of Bible School teaching would now be shelved. But much to my delight I was asked to teach once a week in Northwest Bible School, (now Vanguard College), in Edmonton. This was a trip of 70 miles, but I often hitched a ride with another church commuter. I'm often surprised by where graduates end up serving. Some years later, while attending Nanoose Bay Camp on Vancouver Island, I discovered that nearly all the camp youth leaders had studied under me years earlier at this Bible School. (It's a miracle that they still wanted to be pastors!)

Our children made many long-term friends. Stan was chosen as graduation emcee, despite his recent arrival. Ellen started an "Inter-School Christian

Fellowship Club," with Steve as the President. We began a Children's Church after-school ministry that won new children to Jesus. And Ellen began a younger Women's Group to provide studies for mostly new converts. I started several mid-week cell group Bible studies that increased our Bible study attendance five-fold. People were inviting their neighbours in this large parish area to study in their homes. Every new visitor to our church was invited to eat at a home. The women's group organized several host-volunteer families, and we'd rotate these weekly. Our youth group was the largest in town. We were thrilled to see new people attending, but others were unhappy. "The children are dirtying the rugs! Why another women's group? The youth are having too many fun times!" The church cycle of complaints was beginning once again.

But there were many positive aspects that we enjoyed. I was elected presbyter of the area and also appointed as the provincial youth/executive representative. This involved some travel and retreat attendance. Once we brought our children, who went skiing while we deliberated district business in a hotel close by the Rockies. Another "plus" was the joy of hunting and fishing with a licensed guide who attended the church. The boys trapped beavers (and a skunk)! We shot deer, many ducks and a few geese to feed our family and visitors.

Hunting provided a much-needed relief from the pressures of pastoral life. One summer while we were vacationing with family, Steve stayed home. "Dad," he complained, "It took a week before the phone stopped ringing!" But hunting had some unique perils. Once, while duck hunting, I tried to retrieve a dead duck in a deep, muddy slough. I began sinking in the mud and water got into my chest-deep waders. People have drowned that way, but I made it out (with the dead duck)!

On another hunting trip, Steve and I split up on a small back road. The skies were overcast, and the land mostly flat. It's easy to get lost in those conditions. I got lost and fired my rifle 3 times in quick succession, the universal "distress" signal. Steve answered back with 3 shots, but I soon lost track of directions again. Steve asked nearby hunters to help; "I don't think Dad meant to ditch me here" he told them.

One of the hunters he talked to suddenly saw a tree that seemed ablaze with light. Puzzled, since the skies were overcast, he commented to his friend, "I wonder if the good Lord is giving us a sign?" He gave a shout, and I shouted back, "Please stop, I'm lost!" "Yes, we know, we're looking for you", he replied. The Sunday morning after I told that story I found some little gifts the youth brought me on the pulpit. A compass and a cow-bell!

Folks would bring milk, eggs, chickens and produce from their farms. We needed all the food we could get, because youth would often "crash" overnight at our house. Several youth from un-churched homes asked plaintively, "Can I call you 'Dad and Mom?'" On our kitchen wall we hung a hand-painted sign with this poem:

Guest, you are welcome here, come take your ease.

Get up when you're ready, go to bed when you please.

You don't have to thank us, or laugh at our jokes.

Sit deep and come often, you're one of the folks.

As we had done through the years, we used Christian hospitality to heal hurting souls and win people to Jesus. This did result in a MUCH higher food bill of over $800 per month (in 1983). Despite having a congregation with 190 on the church pictorial directory, the monthly salary was barely $100 more than the Alberta district's minimum stipend to a home mission church pastor. I asked for some additional help. With some reluctance, I was given a car allowance and Ellen was hired as a part-time secretary. Traditionally, pastor's wives have served "for free," but in today's economy many choose to work for wages at an outside job. Every church has different capacities and customs, but in my observation and travels I've found that many pastors' wives make excellent church secretaries or co-pastors.

Ministry Meditation:

"Use hospitality one to another without grudging" (I Peter 4:9). This is a biblical command, repeated by Paul to his helpers, Timothy and Titus. I don't believe we can just shrug and say "That's not my gifting." It was a condition

Paul required in the appointment of bishops. It is both a way of winning souls and assimilating new folks into the church family. Some cite statistics which say, "If a visitor does not form a friendship with someone in church within six weeks, they'll not stay in that church." Are you helping your church to grow?

Other beneficiaries of hospitality are the children. I remember listening in rapt attention to missionaries, pastors and evangelists who stayed or ate at our home. It's one of the reasons I became a missionary; they were held in high esteem in our home. Also, hospitality in Christian homes provides a "safe place" for youth to congregate and build wholesome friendships. Make no mistake about it; your child WILL make friends. If you are hospitable, the chances are much improved that these will be "good friends."

An important part of a pastor's work is to ensure the spiritual health of every person in the congregation under his care. God charged his prophet, Jeremiah: "I have set thee a watchman unto the house of Israel; therefore thou shalt hear the word at my mouth, and warn them...if thou dost not speak to warn the wicked...his blood will I require at thy hand" (Jeremiah 33:7,8). Early one Sunday morning I was awakened to pray. I had a vision of people walking on "the high road" of a raised sidewalk. Below, mired in the mud, were needy sinners. Some were starting up the stairs of salvation. Others on the raised platform, were looking down, tempted to join those in the mud. With tears, I shared the vision with the congregation and pleaded, "Take the high road!"

I was involved in an ongoing investigation of a few members who gave "the appearance of evil." Several witnesses and the board were being quietly consulted. Alas, they could not keep confidences and "spilled the beans" to others. Soon there was a backlash organized by sympathizers. A church vote of confidence was held and I got a substantial majority. But those "who always got their way" were implacable, and began circulating a petition against me. We consulted with our children and Joy said, "Let's split!"

Stan was already in Bible School. I had been reading a book by evangelist C.M. Ward in which he said,"You can lie on a bed of nails and endure suffering like an Indian mystic. Or you can say 'Enough is enough!' and move on. The church is like a large ship that needs tugs to bring it into a safe harbour. Do your part to tug them in, and then let another tug take over before you burn out."

We resigned.

Striving for a pure church took its toll. I once asked the partially reluctant board, "Do you want a church full of warm bodies or a church going to heaven and showing a good example?" At the church business meeting, when complaints began about peripheral issues, I lamented, "My oh my, what a track record this church has!" One board member later took me to task for that statement. I replied, "I've talked to five of your six previous pastors who all told me about how this church has operated. I stand by my statement." I also warned them, "If you keep on doing this, one day you'll have a big church division on your hands." I'd been asked by several people to stay on and start another church nearby.

To my sorrow, several years later the church did have a split, and many of "those who always got their way" left. After some more years, the church that once had **190 people in the directory was closed.** Yet there was a bright point as we prepared to leave and travel in evangelism. We counted **more souls saved during the last ten months of our ministry there than in the previous ten years.** Despite my well-founded fears of retribution for speaking to people about living exemplary Christian lives, God was faithful to honour His Word. Some Sunday mornings I'd ask the board, "Please come and pray for me, the message today is going to be hard to preach!" God would then anoint me with boldness and frequent prophetic words or new songs in the Spirit. Later, people maybe wished they hadn't prayed so hard!

Ministry Meditation:

What is the "anointing of the Spirit" and how do you receive it? Here are some things I've learned that may be helpful.

1. ***ASPIRE to have the anointing, which STARTS in some measure when you receive the baptism in the Holy Spirit and speak in tongues.***

2. **KEEP SEEKING** *this empowering as you minister. A missionary friend, Vic Hedman, told me, "I was asked to preach in place of someone on a Bible School assignment. Almost all the way, at the back of the bus, I kept praying, 'Jesus, anoint me with your Spirit as you were anointed' and God did just that from that day on." Vic also prays a lot daily.*

3. **GET GOD'S MESSAGE** *before you speak with anointing and authority. Spend enough time seeking God's face till you hear from Him. Don't launch out in personal tirades.*

4. *"**BE YE CLEAN** that bear the vessels of the Lord" (Isaiah 52:11). Sin will eventually cut your connection with God. Avoid bitterness, the fountain that contaminates (James 3). Continue to forgive others and return good for evil, even as you plead for their souls.*

5. **FASTING** *and prayer can increase the anointing for special challenges. One missionary, Rev. Adams, said, "It seems that whenever I prayed and fasted, more people were filled with the Spirit in special meetings."*

6. **BE WORKING** *for Jesus with the power He gives. Just seeking "goose bumps"*

often leads to fanaticism and an endless following of evangelists for "another fix."

7. *CHRISTIAN MENTORS can help model successful ministries you long to possess.*

8. *MAINTAIN YOUR FOCUS to be like Jesus and work for eternal goals.*

9. *EXPECT OPPOSITION, "the YOKE shall be DESTROYED, because of the anointing" (Isaiah 10:27). "Christ" was the "anointed one" and Anti-Christ hates the anointing that breaks Satanic bondages. In Argentina I would sometimes see demonic manifestations when I was preaching under a strong anointing. Satan is fearful of "the POWER that is working in us." Thank God for "The victory that overcometh the world" (I John 5:4). "Greater is He that is in you, than he that is in the world" (I John 4:4).*

There are also some things to AVOID in this realm of God's anointing.

1. *DON'T CRITICIZE the behaviour of those who enjoy God's anointing. David's wife, Michal was struck barren in punishment for her contempt.*

2. DON'T SUBSTITUTE the anointing for a deep study of "the Word that abides forever."

3. DON'T FAKE IT, by loud shouting or similar manifestations. You can't fool God, and perceptive people may become cynical.

4. **FEAR GOD, NOT MAN** as you preach with power. Better to offend man than God.

5. **KEEP HUMBLE** (and worthy of being used), by giving all the glory to God.

6. **TAKE A REST**, lest you get burned out by your passion. Evan Roberts, leader of the famous Welsh Revival in 1904-05 had a nervous collapse due mostly to hardly ever sleeping. I believe his ministry never recovered. Jesus advised his disciples to "Come apart and rest a while" (Mark 6:31). "Come apart or you'll 'come-apart.'"

Chapter 10

Starting a Spanish Church in Saskatoon, and Revival in Argentina (1985-90)

Yes, we had resigned the church, but now what to do? It was Christmas and all the year-end celebrations were tinged with sadness. Many new converts were ready to protest and leave the church. I'd started a "new membership" drive, and some of the "old guard" fearing their power would erode further, passed around the above-mentioned petition. They attempted an "end run" around the congregational vote in our favour. But I told the protesters, "Use ballots, not bullets. Stay on, become members, and change the way things have been run around here."

Some years earlier we saw the dramatic play "Heaven's Gates and Hell's Flames." We bought a sound track. I called pastors we knew telling we'd put on this play and would preach for salvation and the baptism in the Spirit. God gave us favour, and we soon filled our schedule for six months. The church kindly let us stay in the parsonage for January and February. Some board members were chastened by the folk for their complicity in ousting yet another pastor. Despite my counsel, many new converts went to other churches or, disillusioned, faded back into their former ways. Words of Jesus about "offending one of these little ones" and "millstones" come to mind, but we let God be the final judge. Sadly, many among the "opposition" lost their children to the world.

We moved our furniture to an empty country church beside a parishioner's home, and set out in our large van for Birch Hills, Sask. and our first week of meetings. As often happens when a missionary or minister travels, we become counsellors to troubled pastors. In place after place, from Saskatchewan to B.C., we saw souls saved after the drama. Joy and Steve sang with a strong anointing and people were filled with the Spirit.

Steve's high school teachers and the school administration were very upset with Steve being pulled out of school just months before grade 12 graduation. "Don't your parents love you? Why are they doing this to you?" they protested. But for Steve and Joy, going back to studying by correspondence was no big deal. They'd study late into the night after our meetings, sleep in, and then take some time off before preparing the "set" for the nightly practices at church. This was a fantastic mentoring experience for them, and they both eventually entered pastoral ministry with confidence. At the end of the school year, Steve hitch-hiked back to his previous school and escorted his former classmate, a nice Christian Egyptian girl, down the aisle for graduation.

My daily evangelism routine was: a) confirm meetings and consult pastors in the morning b) pray most of the afternoon for the meeting c) preach every night d) practise the drama after the meeting. There was very little "spare time." Over the years I suppose I've exercised at different stages, the "five-fold ministry." "And he gave some, apostles (=missionaries/sent ones); and some, prophets; and some, evangelists; and some, pastors and teachers" (Ephesians 4:11). For me, being an evangelist was by far the most stressful, though Ellen "had a lark" being relieved of "pastor's wife" duties. At the end of our months on evangelistic tour, I was diagnosed with high blood pressure.

Once again, we were invited to teach missions in the Saskatoon Central Bible School. We rented a house near the Bible School, instead of commuting from our Dundurn house. Both Stan and Steve entered the same Bible School and later Joy did as well. While we evangelized, Joy sold some of our sermon tapes and Bible stories, from which she got a share of the profits. Steve brought our vacuum and carpet cleaning equipment. He'd stand up and say "I'm going to Bible School, but I've got to pay for it. So please ask me to clean your carpet. If it's not dirty bring the dog in, or throw a party!" I'd subsidized the first year Bible School fees for Stan and Joy, but Steve had enough to pay for the whole year of studies from his cleaning jobs.

Ministry Meditation:

Evangelists are one of God's gifts to the church. They can help a pastor reap a harvest from his sowing the good seed of the Word. But how do evangelists make a living? The short answer is "with great difficulty." If it had not been for some residual unemployment insurance, and having no rental payments, we would not have made it. Years ago, Gordon Townsend, a B.C. pastor, was challenged to "do the work of the evangelist" and see first hand the difficulties they faced. After travelling for some time, including ministry in Nelson that I remember as a child, this pastor/evangelist/author wrote, "How to Treat the Church Evangelist." Over 60 years later, the book's advice is still valid, (and is still not being heeded).

Consider the following facts. You are fortunate to get 35 weeks of ministry each year. (Think "down times" like summer holidays, Christmas, Easter, hard winters). That leaves many weeks with no meetings and no income. You have no expense account, car allowance, parsonage or house allowance and often must maintain a home for the family. (Some travel with the entire family in a large trailer, but trailers aren't cheap either nor practical in winter). And God forbid that you have a sudden cancellation close to the date of a scheduled meeting! You can't arrange another meeting on short notice, and boards rarely consider paying any compensation.

Pastors sometimes say, "This is what I get paid each week (and that's what we're paying you)." Well yes, but don't you get vacation paid time? Utilities, car allowance, Christmas gifts, office phone and supplies, income while you are sick, conference expenses and many other costs are all "additional incomes" a pastor receives that most evangelists pay 'out of pocket'. Bottom line – an evangelist needs about twice what a full time pastor is paid to compensate for his costs. Based on my experience, our church board set a new evangelist mileage travel allowance policy, plus a standard compensation that could be increased by any love offering exceeding this amount. We had some very grateful evangelists thank us for this.

It was nice to have all our children "home" with us once again. A good education will eventually pay for itself. And when a student has to pay for his own studies, that wonderfully sharpens the mind. We had very little income, but our children now were helping us. We even "borrowed" from their student loans (which we paid back later). After a few years we were then able to help them when they got married.

Stan found a nice girl from Ottawa called "Sandy Dempster" in Bible School. This was following a familiar pattern. Sometimes while promoting Bible

School, I'd say, "And another benefit is that you might find your wife. We were four Anonby boys who all went to Bible School and three found their wives there. We had three children, and two found their spouses in Bible School!"

I was "burned out" for pastoral ministry at this time. **"Burnout" often happens in the "helping professions." The main elements are being over-worked and under-appreciated or under-paid.** We did weekend ministry in nearby churches, and once again trained all our children and Bible School students to perform "Heaven's Gates and Hell's Flames." This time we memorized the lines, instead of "miming" with a background tape. Steve and Joy were the main organizers, putting together the mobile "set" I'd made. These were wonderful meetings, but not enough to pay our bills. I tried to start a commercial cleaning business, but the market was saturated with Spanish immigrants.

Ellen got a part time job with a house-cleaning business and suggested we try that instead of large stores. I produced some ads and we walked through nearly 40% of Saskatoon, with its 240,000+ inhabitants, delivering them door to door. We soon had enough contracts for both of us, and then we needed employees. One Sunday morning we walked into Westside Pentecostal Church and met a couple from El Salvador. When she heard we were Spanish-speaking missionaries, Mrs. Quintanilla burst into tears. "We've been looking for some Spanish pastors to teach us the Bible!" she cried.

We promptly started home Bible studies and the group began growing and growing till we were holding meetings on both sides of the N. Saskatchewan River that bisects Saskatoon. Many immigrants told horror stories of political repression. Mr. Quintanilla found his father's head impaled on a stick at the bottom of a dry well. Another man reported fraud and was forced into hiding, fleeing from one "safe house" to another. Even his wife was not permitted to know his whereabouts. These people fled for safety to the United States, but their residence applications were dismissed as being "economic refugees." In mortal fear of being killed if they returned to these (mostly) Central American countries, they fled again to Canada. There they found a more sympathetic hearing and were sent to various target cities across Canada.

Ellen and I would meet them at the airport, give them cookies, and invite them to our home meetings. I was reluctant to become a pastor again, but these needy folks begged us to pastor them. The Westside Church, under pastor Moffat, invited us to use their facilities on Sunday afternoons. About half of the group were not Christians; they just liked to get together and meet Spanish people.

My son Stan co-pastored the church with me and took care of a growing youth group. The Lord had brought me back to the mission field – in Canada. We had 12 Spanish nations represented in our meetings. It was time to form a "proper church", and we did so, under the generous covering of Westside Church.

Our house and office-cleaning business became very busy, with close to 50 houses being serviced monthly. Our new Spanish friends became our chief employees. They were, for the most part, well-educated and skillful. We used our large "work van" as a "church van" on Sundays. Many would come to church on Sundays for "payday," and eventually found the Pearl of Great Price, Jesus, as their personal Saviour.

It was now 1988 and we'd been back in Canada already eight years, instead of the five years I'd estimated. This was our 25th wedding anniversary year in the month of October. Our daughter-in-law, Sandy, wanted to gather donations for us to take a celebration trip to Hawaii. "No, no, I want to go back to Argentina!" Ellen insisted. My folks and two of my brothers came from Vancouver for our 25th celebration.

Afterwards, we took the plane leaving Saskatchewan's Centigrade -37 (-35F) below temps and landed in Argentina's +37 (98F) degrees above temps. I peeled off my shirt to swim and get cool. But I got so much sun exposure that my white skin was severely burned and I peeled like a growing snake. We began in Mendoza, and preached 15 times in 30 days as we travelled all the way north to Salta and Jujuy. Everywhere, churches were revived and growing after a national revival spear-headed by Carlos Anacondia.

Carlos had been a businessman, but experienced a miraculous salvation and healing. God gave him favour with all the evangelicals and he filled stadiums with crowds excited to get a touch from God. This move began to extend all over the country and abroad. Many ministers began to believe God for greater manifestations of His power.

We crept unannounced into our Salta church on a prayer-meeting night. We'd left a congregation of about 40. Now there were over 250 at a prayer meeting! The man who led in songs was a doctor. His wife had been preparing for baptism and he was not about to be left behind. He gave his heart to the Lord and joined her in the ceremony. Some months before he'd been depressed because he'd recently graduated as a doctor but no hospital would accept him. He poured out his woes to a Christian taxi driver. The driver said "You'll have a job by next week! I'll get the church praying about this!"

Sure enough, he was granted directorship of a hospital within a week. It was filled with "hard cases," including one man with incurable running sores. He gave him a prescription that said, "Go and get prayed for by a Christian pastor." Soon this "hard case" was cured and the doctor kept getting "hopeless cases" sent to him, that only God could cure.

An especially dramatic healing was recounted to me in Salta by a Spirit-filled Church of God missionary. An army general's wife was on her deathbed in the hospital near where we used to live. The general went to the site of a recent campaign, because "everyone knows that evangelicals pray for the sick."

"Sorry" they told him, "Evangelist Clements has finished the campaign and is taking the plane back to Buenos Aires." The general jumped into his car, raced to the airport, drove down the runway and stopped the plane. (You can only do that if you're a desperate general)! Meanwhile a frightened stewardess repeated this announcement: "Will Dr. Clements please identify himself?"

"There must be some mistake, I'm pastor Clements, not Dr. Clements" he explained. In Argentina, due to pressure from doctors and the Church, those who lay hands on the sick can be charged with "practising medicine without a licence." It was a nervous moment, but the stewardess said, "It's you we want!"

The general shoved him into the car and sped back to the hospital. **His wife was now dead, but God raised her up from the dead!** The general declared, "I will open doors for you among important people." One day Pastor Clements was invited to a political meeting. When he was introduced to a former governor of the small province of La Rioja, the Spirit of God came upon him. "Congratulations!" he said, "You are going to be the next President of Argentina!"

The governor was taken aback. "I'm not even thinking of running!" At that time the evangelical church in Argentina was fearful that the government and official church would take steps to squash the recent revival. But governor Carlos Mennon became President,and was re-elected for an unprecedented second term. Having been miraculously proclaimed as a future President by an evangelical pastor's "word of knowledge," he dared not touch this revival.

I got an important prophetic word myself in Pastor Jose Vena's large new Buenos Aires church. We'd started the Salta church with pastor Vena seven years earlier. He came chronically late to meetings, and spent a lot of time sleeping. We almost despaired about him succeeding at pastoring. Now, after we'd ministered, prophetic words were spoken to me. I was told to heed the words of God to a complaining Jeremiah. "If thou return, then will I bring thee again, and thou shalt stand before me: and if thou take forth the precious from the vile, thou shalt be as my mouth" (Jeremiah 15:19). The Lord had seen bitterness in my spirit at mistreatment in our former church and at the hands of leadership who knew of past problems, but had not warned me, nor were they very sympathetic. I'd felt abandoned and suffered financially for being a "faithful prophet," It wasn't fair!...but God couldn't use me like that. No wonder my application to return to the mission field kept getting delayed!

The Lord had another "word" for me. Jeremiah is again reproved by God. He'd been grumbling, "wherefore doth the way of the wicked prosper?" (Jeremiah 12:1). God replies "If thou hast run with the footmen, and they have wearied thee, then how canst thou contend with horses? And ...how wilt thou do in the swelling of Jordan?" (Jeremiah 12:5). In other words, "Buck up and endure hardness as a soldier of Jesus Christ" (2 Timothy 2:3). The Lord healed my heart and drained the "swamp" of resentments. It wouldn't be long now before we returned to the mission field–this time in Spain. I could inject a "ministry meditation" here but I think the lesson is clear.

Now our reappointment as missionaries began to gather momentum. Stanley moved with his family and our new grand-daughter, Jillian, to Prince George. His wife, Sandy, was accepted into the nursing program and Stanley began teaching in the Northern B.C. College. Steve and Joy transferred to our old "alma mater," now called "Summit Pacific College," and relocated in Abbotsford, B.C.

Now we needed a new youth pastor. Lorena Navarette, one of my Bible School students orginally from El Salvador, took that position and soon became the Spanish church pastor. We were able to get some home mission department financial support for our new church. We left the church in Lorena's "good hands" and, after a brief itineration in B.C., headed as missionaries to Spain. It had been a long ten years, but our eldest child was now married, and Joy married Ken Kutney, a pastor's son, while attending Bible School. The "re-entry to Canadian culture" for our children now was "mission accomplished."

Chapter 11

Missionaries to Barcelona, Spain (1990-93)

Our ten year "interlude" from overseas missions work was over. Joy lamented, "Children leave their parents, but my parents left me!" But truthfully our family was happy to see us back in the missionary ministry they knew we longed for. In an age of working telephones, (at least in Spain), and the soon-to-be internet age, communication distances have shrunk drastically.

Barcelona is the second-largest city, (over four million), in Spain. It is also the capital city of the province of Cataluña, an area that has many speakers of "Catalan," a language that is an amalgam of Italian, Portuguese, Spanish and French. 800 years ago it was a part of Charlamagne's empire and they still glory in that fact. They have some quaint group dances and like to build "people towers" that rise up to seven layers. Barcelona also boasts a beautiful tree-lined "rambla" (boulevard) and the famous "sagrada familia" (sacred family) cathedral designed by Gaudi. (It looks so odd that the word "gaudy" derives from the design).

Our first assigned task was to minister in a tiny church in Barcelona. The church began as an outreach by Rev. Vidal, father of the now-famous singer/composer, Marcos Vidal. We heard they had a congregation of 16, then 14, then 12. By the time we arrived it was about eight regulars. A deacon was doing his best to shepherd them and we alternated ministry days. I didn't want to simply shove him aside after his faithful work. The Lord had it all arranged. The deacon got very ill and conceded the church to my care. He later recovered completely.

We wondered why the church had kept shrinking. I had been told "Gypsies and Spanish don't mix!" However our compassionate God began sending a number of people "from the other side of the tracks." Some curious former drug addicts attended and loved the sense of God's presence. Word spread and attendance kept climbing.

One of the new attenders, Laura, was married to Juan, a bar owner. In Spain, most people live in apartments, and nearly every downtown square block has at least a couple of bars. They also legally serve alcohol at the price of a soda to 16 year olds in MacDonalds. What is the result of early exposure to alcohol in the home? Close to 1/3 of men over 40 are alcoholics. And I never met one drug addict who didn't have at least one alcoholic parent. Laura noticed that her husband behaved better when he was going to church and professed salvation. As she saw him slipping, she decided to go to church herself. Soon she was saved, filled with the Spirit, and brought her husband back to church.

Juan began witnessing to people in his bar and asked us to go with him to jail to visit his old friends. He'd tell drunks, "Don't drink so much, come to church and be delivered!" He brought along a fine-looking drunk who was a homosexual. God delivered him wonderfully, and being a book-binder, he rebound my old Bible for free. Juan opened his back room in the bar for meetings. Can you imagine the opportunity with sinners so close at hand?

Andres, a Gypsy, came to church slightly tipsy. He was our very first visitor in many weeks. At one point he stood up to say, "The Bible I got from the Christian radio station is different!" He was told, "Sit down and shut up!" He walked out, as Ellen ran after him saying, "Stay, we love you!" Some weeks later he returned, this time with his substantially-sized Mom. He sat through the service quietly, and in due time both he and his mother accepted Jesus. He still had some struggles with his addiction to strong drink, but he was a witty character who loved to talk. Talking in church brought new problems to the formerly quiet church.

One Sunday morning we heard a sudden loud thumping on the piano. "Sit down and be quiet, show respect to the house of the Lord!" The deacon's daughter (and pianist) had all she could take with these new attenders. Her mother refused to shake hands with Andres and other former drug addicts. In a small church, such slights are easily noticed. The church custom was to enter and pray quietly, listen to the sermon and only afterwards greet others. Sometimes a church sign at the front told you to do just that. The new attenders froze and next Sunday came late. After some months, with the church now packed full, the deacon's family and some "die-hards" left the church. No wonder it had been shrinking!

The new converts shrugged and said, "We knew they never loved us!"

Fast forward some years, and we returned to the church on a surprise visit. Andres' mom greeted us with the news that "Andres doesn't drink any more." Andres came by and said "I don't drink any more!" The former homosexual was freed from his bondage. I'd been challenged by some of the "old guard" with the question, "If these new people are really saved, why don't they act like us?" I wasn't sure that would be a big improvement!

Ministry meditation:

At the time of this influx of new converts from "rough" backgrounds, I was reading a book written by a former British witch. She described a terrible lifestyle of depravity. Parties celebrating Satanic rituals were unrestrained orgies exalting a defiance to all things good. One night this group was celebrating rituals naked on a beach. They saw an approaching group of people and were frightened. The witch said "Don't worry" and MADE

THEM ALL INVISIBLE while the visitors walked right past them.

Yet, strong as the witch's power was, God's love was stronger yet. She listened to the witness of a Christian and prayed for salvation. For many weeks, while professing Christ, SHE STILL MADE HER LIVING SELLING HER FAVOURS. It took time, but finally she was re-made in the image of Christ. I believe we need to trust the Holy Spirit's work in making people holy. They may not adopt all our "Christian customs." However, "Man looketh on the outward appearance, but the Lord looketh on the heart" (I Samuel 16:7).

Some studies show that new Christians win more souls to Christ in their first two years of serving Christ than in all the rest of their lives. In part, it is because they have not made such extreme changes in their outward appearance that they look "too different." They don't "burn out" their old friend circle, but instead serve as an intriguing new example. I've sadly seen some older Christians get so demanding of changes that new converts are discouraged and leave. I wonder if this, in part, can be "offending one of these little ones." Patience, prayer, love, hospitality and the modelling that goes on in a Christian home can be a powerful unspoken example. Let the Holy Spirit make people holy. Keep the faith. God will do the work all in His good time.

In 1992, Spain's 500 year anniversary of "discovering" America, they were the host nation for both the Olympics (in Barcelona) and the World Expo (in Seville). We hosted over 250 Canadian youth who worked hard to evangelize in the Barcelona area. It was hard to get government permits for meetings, but a bold Youth with a Mission group did some powerful dramas in a city park. One of the onlookers said, "I'm in charge of all the city parks. I like what you're doing; you have my permission to hold these meetings on our plazas!" YWAM then extended their permit to cover all other Christian groups.

We got a group that put on the powerful drama, "Heaven's Gates and Hell's Flames," which we translated into Spanish on a background tape. We were able to perform in a large outdoor arena with hundreds listening from apartment buildings nearby. A delegation from the small denomination with which we were working was visiting us. We were astonished that they were offended by not being housed in a Five Star hotel. All our money was going into this campaign, but they saw the drama as "too big for us, just too much!" They were further offended that a missionary whom they had fired (for lack of "obedience"), had returned with the TV program "100 Huntley Street" to lead a Christian outreach in Seville.

There the gospel was shared in song, drama and testimony. The archbishop of Seville attended twice and was moved to say, "I just didn't realize the gospel was so simple!" Many thousands gave their hearts to Jesus and large amounts of Christian literature were distributed. The local denomination, whose leader was under a mental strain, then threatened to fire us and the other remaining missionary in revenge. His crime? He had dared to briefly return to Spain with another para-church organization.

At that time, the PAOC was "down-sizing" its missionary personnel due to severe financial problems. They had built a new head office building and the deal to sell the old building fell through. They were planning to shut five mission fields. I was told, "We only can afford one missionary." What to do? At the age of 50, with all my life spent in the same denomination, where was I to go now? And who had really placed the call to missions on our lives?

Right then, in our confusion and sorrow, we were invited to a retreat especially for missionaries working in Spain. "Operation Barnabas," as it was called, sponsored a free retreat for three days to all who wanted to come. My distant cousin in N. Spain invited us to come. The drop-out rate of missionaries to Spain was nearly 100% after ten years. The conversion rate was very low – about 120,000 evangelicals, less than .03% of the population. Half of these were Gypsies, the so-called "Hallelujahs."

Despite the announcement that the retreat had a "Charismatic flavour," only two attending families were Pentecostal. The burning question around the dinner tables was "Having any results, any converts?" I was one of the very few who said, "Yes, praise the Lord, our church is packed out!" I recall a couple formerly from Columbia who quizzed me, "How?"

I looked at the tight-lipped husband and decided he would not approve of what I was about to say. So I temporized, "Well, people don't like to come into an evangelical church in Spain, so we had barbecues and lots of social times in our house and yard. Soon we'd sing choruses, make friends, testify and then they'd come to church." I could sense the lady was taking mental notes. But the next morning my conscience smote me. I sought out the lady and said,"Well, the **real story was the powerful presence of God anointing the messages and then the Spirit fell!** When folk were moved by the palpable conviction of God's Spirit, they repented and their lives were changed. So, my friend, **that's why the church grew!**" She smiled kindly and thanked me.

At this time the Pentecostal Assemblies asked us to come back to Canada with no promise of appointment elsewhere. Our former missions director, Rev. Cornelius, was scheduled to meet with the small Spanish denomination to work out a new agreement. Another leader had come and suggested we work with another organization, or just move out entirely. Cornelius hated the constant conflict. He had a brain hemorrhage and died just days before the scheduled meeting.

We were in a quandary. We had begun a church in a different area of Barcelona and needed to get a new leader. Also, a renewed visa was about to be issued. We were looking for another organization with which to continue as missionaries, and the Assemblies of God had asked us to consider joining them. We did not feel we could simply abandon this new work, and asked for more time to finish the work. New missions leadership in Canada faxed me a letter which said, in essence, "since you are not following our instructions to return at this date, we will not pay your return fare home, and your salary is suspended!"

Now, what were we going to do?

God opened a door for us to live in a Youth with a Mission (YWAM) hospitality center and serve as hosts for several months. An independent church in Canada paid us $1000 per month for six months. Our visa to return to Spain was granted and we left a few days later. Both churches now had their pastors in place and we could (and did) tell groups that had helped us, that "the fruit was preserved."

Still my heart grieved for the many souls yet unreached. And once again the Lord spoke to me through His encouraging word: "**My word shall not return unto me void** but it shall accomplish that which I please, and it shall prosper in the thing whereto I sent it. For **ye shall go out with JOY and be led forth with peace**" (Isaiah 55: 11, 12).

These words brought assurance and peace to our hearts. We were not wilful rebels, but just wanted the work to be left in good hands, and not to waste all our recent "Olympic effort." A few months later, the PAOC gave me an unexpected $5000 additional settlement and a gracious letter of transfer to the Assemblies of God. It took a few years, but one of the leaders of the Spanish denomination apologized profusely to us for their attempts to expel us missionaries. "Who were we to close the doors to someone wanting to win souls, obey the Great Commission, and start churches?"

Ministry Meditation:

This was not the first nor the last time I've observed tiny overseas denominations attempting to legislate who can evangelize in their country. Leaders can become so anxious to retain their positions that they fear any competition that might show them up. Missionaries attempt to cooperate with any existing full gospel groups whenever possible. But it is unthinkable to limit labourers when the harvest is so pitifully small. The fishing disciples called for more boats to help bring in the great catch of fishes.

A parallel problem occurred throughout the history of the Christian church. The Pharisees persecuted the Christians, and astute Pilate "knew that for ENVY they had delivered him" (Matthew 27:18). "...behold the world is gone after him" (John 12:19). One after another, church history shows how the Roman church persecuted the Lutherans, the Lutherans the Reformists, England's Anglican church the Methodists and 'free churches', traditional churches in America the Pentecostals, and some Pentecostals others who have "too radical" ideas of government or practice.

I am not defending deliberate and calculated church-splitting to start a new church, nor "building on another man's foundation" to kick-start a new denomination. But a certain amount of division and competition helps keep a church or an entire denomination "on its toes." For example, at least a part of the Charismatic movement in the Roman Catholic church was an attempt to sincerely seek the miraculous power of God when they saw people leaving their church. Luther had hoped to reform the church, but was instead expelled and reluctantly had to leave it.

We need to be cautious lest we find ourselves "fighting against God" (Acts 5:39). We must contend for Bible truth, but our attitude is important. Paul advises Timothy to demonstrate "MEEKNESS when instructing those who oppose themselves; if God peradventure will give them repentance to the acknowledging of the truth" (2 Timothy 2:25). God deplores His children fighting and the world scoffs at a divided church. Absolutely nobody can claim they have the complete truth. But LOVE is eternal and the trade mark of true Christians "by THIS shall all men know that ye are my disciples" (John 13:35).

The same divisive spirit often occurs at the local church level. It may be due to long-standing power struggles. We hear of it in 2 John, verse 9: "Diotrephes, who loveth to have the preeminence among them, receiveth us not." Some people actually enjoy conflicts as a spectator sport. But God numbers "he that soweth discord among the brethren" (Proverbs 6:19), as one of the seven abominations that He hates.

Most often I've observed church-splitters and dissidents as justifying their often underhanded stratagems and criticisms under a cloak of spirituality. They've got the "new truth" and those who don't follow them are are spiritual simpletons or enemies. They usually experience several unexpected consequences, however.

First, new believers become disillusioned and leave the church as they are forced to take sides against former friends. Second, the church can go into a downward spiral as its reputation is besmirched in the town. Thirdly, their own children are confused when they have "roast pastor" for Sunday dinner. How can they be fed spiritual counsel by a pastor whom their parents disparage? If their parents leave a church, they lose friends, especially when this happens more than once. If Mom and Dad "get off the bus" once too often, children may say "That's it, my parents are nuts and I'm not going to the next church!"

I once had a frightening vision that still brings tears to my eyes as I write. In this vision I see a heart-sick Christian giving up. "That's it, if that's the way Christians are going to be, I'm through with God!" At that moment, as his back is turned to go, the offending person peels off his mask, revealing a Satanic face. "I fooled them again!" he gloats, "I fooled them again!" Satan has his allies, ("moles" in spy terminology), who are "sleeper agents" in many churches. They may look and act like

Christians gaining acceptance to the point they may even consider themselves as "Christians." Hypocrites WANT to appear better than they are. But one day God will say "I NEVER KNEW YOU: depart from me, ye that work iniquity" (Matthew 7:23). Be careful and prayerful. When Jesus said,"One of you will betray me" all the disciples asked, "Is it I?" I believe each one saw their vulnerability. In short order they denied Christ and fled.

This was a sobering revelation to me, forged on the anvil of personal experiences and the stories of others who have gone through devastating disappointments. From time to time I hear of God giving a word of knowledge or a gift of discernment of spirits that exposes deception in the church. Make no mistake about it; we are in a "dirty war" with a foe that will use any tool and any fool to do his work. Army generals of past generations used to prepare for battle by playing the military game of chess. The "front men" in a brazen frontal attack are called "pawns and cannon fodder." They are expendable. Don't be a simpleton in someone's game to attain power in a church fight. Keep praying instead of plotting. "Blessed are the PEACEMAKERS: for they shall be called the children of God" (Matthew 5:9).

Castlers up to 7 high

Sacred Family church by Guadi in Barcelona

Orange crop in Southern Spain

Paella feast

Chapter 12

To the USA and the Canary Islands with the Assemblies of God (1994-2001)

Back in Canada again, we visited our son, Stan, in Prince George, set in the forests of northern B.C. While awaiting appointment as Assembly of God missionaries, we began raising support among Canadian family, friends and churches. God gave us favour, especially in these northern churches. Stan had told Pastor Arlo Johnson about our plight in Spain, and he helped us with six months of interim support.

At a Christmas banquet fund-raiser, Pastor Arlo asked me, "How much do you think we should set as a goal?" "One thousand dollars a month" were the words the Lord just put into my mouth. Arlo gulped and said nothing. Pledge forms were placed by each place-setting. Next morning he came to the house very excited, "$1000 exactly was pledged,

$1000 exactly!" And, true to their word, the church held to that amount for five years until there was a change in the pastorate. A member who worked with the government helped me set up a government-certified missionary charity to receive tax deductible donations A five member board now guides our activities and oversees the financial reports since 1994.

After a batch of tests, interviews and a retreat with over 40 candidates, we were approved for missionary service with the Assemblies of God. But nobody said getting a "green card" would be easy, although ministers are supposed to have more ready acceptance. In order to save time while we were waiting, we visited churches in Pennsylvania but received no personal monies. How did we live? Well, my son, Steve was given a used car and sold us a little green Honda Civic for $20. The battery problem turned out to be a minor wire connection, easily fixed. The car "ran on air," and we actually made money on the vehicle mileage reimbursed expenses. Pastor Decker, from the church in Christiana, PA gave us free lodging in the church basement. And with $1000 per month plus some other Canadian donations, we were able to survive while building up our future missions account.

The missions committee in Springfield wondered how I would do as a Canadian raising funds. I was ready for the challenge. I had bought a HUGE "stars and stripes" necktie. "I'll just put on this tie and tell them 'I'm an American'!" I told them. That brought a hearty laugh followed by, "You'll do okay!" We diligently phoned scores of churches and began our fund-raising. As experienced missionaries, we had a clear message to share and a track record of God's blessing. The church in Christiana had been an independent church, birthed in a revival among the Amish. You may have heard of "Auntie Anne's" pretzels. She was saved in this church and became a strong missions supporter.

The church had several missionaries they supported with an unprecedented faith promise of $1000 per month. The missions board met in the hall next to our little windowless evangelist quarters. Ellen baked them a nice pie every month. And would you believe it? They voted to grant us a monthly pledge of $1000 per month! I was pleasantly surprised at how generous the churches were in their missions offerings. Many churches had a goal of giving at least 10% of their income to missions. The way that most did it was by missions conventions with faith promises.

Ministry Meditation:

Oswald Smith was a prolific songwriter ("<u>Then Jesus Came</u>"), writer ("<u>A Passion for Souls</u>"), pastor (People's Church, Toronto, Canada), and missionary evangelist. His health broke twice as he attempted to be a missionary to India. But his church was probably the first in North America to exceed, many years ago, $1 million per year in missions giving. How did he do it? He used the annual "Faith Promise Missions Convention" to set goals and promote giving. Every segment of the church, from

Sunday School classes, choirs, youth, men's and women's groups was involved.

Missionaries and special speakers were brought in from around the world. Costumes, displays, flags, banners, marches and special music were featured. Every one of the 300 elders was expected to participate in giving their faith pledge. The pastor set an example. A large screen showed the progress of incoming pledges. The atmosphere was electric as people prayed the year's goal would be reached. It was a high point of the church year. Throughout the year, the entry of funds was monitored and reported upon. If it was less than expected, there was a call to prayer. Every year the goal was met.

I have been at some USA churches where this yearly Faith Promise Missions Convention is their high point of the year. Without exception, it has boosted their giving dramatically. The church in Roaring Spring, PA, under pastor Tony Baker is a wonderful example. The Lord has blessed their church and they built a large new church without lessening their missions offerings.

But how does a person get the extra money to give to missions? Well, it is a "faith promise" that is made, not a binding legal obligation. I have preached several such conventions and this is my message about how God can help us invest in eternal things, like winning the world for Christ.

FIRST, LIVE SIMPLY. Have at least a rudimentary budget so you know where your money goes. A secular book entitled something like "<u>The $5000 latte</u>" shows how a daily cup of coffee and other such "mini-splurges" amount to "real money" that could be better invested. Some people are "shopaholics" who get emotional relief that way. What can we do to prune some expenses from what economists term "discretionary expenses?" (That is, income that is not ear-marked for basic costs such as housing, food, clothes, taxes, transport, health and miscellaneous expenses). Think moderation in: holidays, cars, houses, furniture, clothes, cafes, education, gifts, entertainment. Living more simply and "down-sizing" can free up a lot of money. The most important person to impress is God, who will ask us to "give an account of our stewardship" (Luke 16:2).

SECOND, RECOGNIZE THAT GOD CHALLENGES THE FAITHFUL GIVER AND TITHER TO "PROVE ME...I WILL POUR YOU OUT A BLESSING" (Malachi 3:10). This is the only verse in the Bible that God challenges his children to "PROVE Him." God WANTS to bless us with abundance, but some may abuse His blessings. So God tells us to demonstrate our faith by continued generosity.

We need to recognize God's abundance in many forms. For example: unexpected gifts of money; food or clothes that we need; a raise in pay or profits from a business; sudden "bargains" and a host of other blessings. Again, if you keep a careful record/budget, you'll be more aware of God's blessings. You can turn some of these over to "laying up treasures in heaven," by giving to your missions faith promise.

THIRD God promises the generous giver that He will "REPROVE THE DEVOURER for your sakes and he shall NOT DESTROY the fruits" (Malachi 3:11). Think of how your earnings or savings can disappear suddenly. In the stock market and housing fall of 2008, many people lost over 40% of their net worth or retirement savings. The list of potential disasters is long: unemployment, fires, litigation, sickness, bankruptcy, devaluation/inflation, – any of these things can devastate our finances. God does not guarantee that we will face no trials, but He helps us escape some of the worst effects.

Quite a few of the "money-loser" disasters are morality-related (think drug addiction, alcoholism, law-breaking, unfettered greed). We were stolen from about 20 times in Argentina, during our first term. I finally adopted the philosophy "Lay up for YOURSELVES treasures in heaven...where thieves do not break through nor steal, for where your treasure is there will your HEART BE also" (Matthew 6:20, 21). You can have extra money by avoiding financial leakages. As a bonus, God has a "lock" on your heart and an eternal reward for you.

We enjoyed our missionary itineration travels in Pennsylvania. Our trusty little green Honda Civic hatchback kept purring along for thousands of miles, visiting over a 100 churches throughout Pennsylvania's forests, farms and factories. We took turns driving, as the other would recline the passenger seat and sleep. We had to shore up the back floor due to rust, but the motor never missed a beat!

Due to spending very little, we had our pledges and missionary account requirements completed in record time. Until a missionary has full promise of support, they are not normally cleared for travel to the field. We came to an agreement with the Assemblies of God Missions Dep't that we would accept 50% of our allowance, and depend on the Anonby Missionary Society in Canada for the balance. Yet we still did not have our Greeen Card, nor an entry visa to Spain. What to do? The area director, Jim Nealy, decided to let me travel on a three month tourist visa and live off contributions from our Society.

We had been slated to teach in the Madrid Bible School. Ellen was apprehensive about living in the unrelenting summer heat. She'd had a heat stroke as a child, which still caused her dizziness if she got too much direct sun. One day, soon before leaving, we got a phone call from Jim Nealy. "There is now a greater need in Spain's Canary Islands. Would you consider changing destinations?" I hung up and put on a sober face as I looked for Ellen. "They asked us to go to another place. Would you consider going to the Canary Islands?" A big whoop and "Whoopie!" was her prompt reaction. She'd been willing to go to hot Madrid and somehow trust God to intervene. Yes, God indeed intervened.

Where are the Canary Islands? They are two Spanish island provinces lying off the coast of N. Africa, (Morocco). It is a two or three hour flight from Spain, crossing the Mediterranean Sea and the nearby Sahara desert. They are comprised of seven main volcanic islands, with a resident population of two million. But every year ten million European tourists come to party. Every island has beaches, tourist hotels and nightclubs.

The Guanches, original inhabitants from Morocco, were displaced, enslaved, killed, or died by suicide as they hurled themselves from high cliffs. Some inter-married and settled down. Yes, there are some wild canaries. But the new Spanish conquerors were afraid of their fierce shepherd dogs, ("canarie"), thus the name "Canary Islands." Laws were passed to limit shepherd dogs to one per flock, which enabled Spanish "rustlers" to steal these flocks. Many people lived in caves, especially the "guanartemes" ("kings"). Cool in summer and warm in winter they were the preferred dwelling. There are scores of these caves that are still fixed up and occupied. Many have a coat of white waterproof paint on the exterior stone ceiling, because lava rock is partially porous.

One of the many tourist attractions in Gran Canaria island is a large cave made into a restaurant. There are many rooms with stone tables and benches carved into the rock face. It is set into a mountain top that has several occupied homes and some smaller cafes. One cave houses a couple of cows, and reminds me of Jesus' birthplace. A local showed us a typical native knife made by his distant grandparent. But, strangely enough, he denied having any Guanche ancestry.

Native warriors rained down rocks and spears on the Spanish invaders. These nimble mountain dwellers used 10-20 foot long vaulting poles to scramble down steep slopes. We saw demonstrations of how they could accurately strike a tiny coin in a crevice with their poles, then quickly slide down the pole and vault to another spot. It took many years but the Spanish finally conquered them by the use of recently-invented muskets.

On the 6000+ feet high uplands grow the Canary pines. From the mountaintop you can see the African coast on a clear morning. From the island of Gran Canaria where we lived, we could see the high and snow-topped crest of 12,198 ft. (3718 M.) "El Teide" on the island of Tenerife. We once took an aerial tram near to the top. The actual flora on the top is protected and only hikers with a permit are allowed there. My determined brother John with my son Stan got those permits and hiked to the top.

The little island of La Gomera is trying to preserve their historic "whistle language" of over 600 words. They used the whistles to communicate from hilltop to hilltop. I have a tape of the island's haunting flute music and whistles. Each island area has a typical costume and their own unique "timple," a tiny guitar. They tend to sing their folklore tunes a bit sharp, and wiggle their jaws to effect a vibrato sound. These islands were the last post of civilization that Columbus visited before crossing the Atlantic Ocean to America.

When we arrived at the Canary Islands, Dr. Mark Barclift was the Assemblies of God Bible School director. There were two other long-time AG missionaries, Joe Szabo and the Mazers. The Mazers were in their 70's, and beginning their last church in Los Cristianos. This was my first "full-time" posting as a missionary teacher, rather than a church pioneer. We had barely settled in when a call came from the American Embassy in Vancouver. Our resident visa had been approved, but we had to appear in person to pick it up.

We returned now as "officially-hired" missionaries of the AG, and were able at last to get paid! It had been a long wait, but God had supplied. One of the unusual ways God provided was by the father of our distant Norwegian relative from the Kasa family, who were serving in N. Spain. She told me, "We'll tell Dad about your need, and you just sit tight." Bro. Billman was a brilliant missionary eye surgeon, with a gruff manner of speaking. After observing us for two days, just before we returned to Barcelona he said, "Come here! You liked my West African shawl, here it is. Here is $100 in cash. And I'll give you a check for $4000, no more." I wasn't going to argue about that! We also got a $10,000 donation from the "Youth With A Mission" couple whose car we'd fixed many years ago. Some months later, the missions personnel director asked me about debts. I answered, "We have more money in the bank than we ever had! It's a miracle!" This was the second time God paid our debts by miracles before we left for the mission field.

We were welcomed by missionary-educator, Dr. Mark Barclift. He had a 12 year-old son who was getting very hard to handle due to a serious health issue. I was left in charge of the Bible School while they went to the States for medical evaluation. I joked, "Make sure you come back!"

Mark's first words upon his return were, "We have to move to the States for continued treatment, Joseph." Wow, I didn't expect to be so quickly called upon to direct a Bible School. The national superintendent, Jose Enrique, was not so sure about that. I'd offended a couple of pastors by speaking too bluntly, and this position required a lot of tact and diplomacy. I give all the credit for learning some diplomacy to my dear wife!

Ellen laid on wonderful meals as we invited, one after another, all the main Church leaders to our home. We would eat, sometimes play a homemade croquet game (with tennis balls), and get to know each other informally. Several non-AG denominations were cooperating in financing and sending their students. The mainland Bible School, in Madrid, with a mainland population of 38 million to draw from, had fewer students than we did. We inherited a school with considerable debts, but at the end of six years we were solvent and had added new facilities for married couples. Ellen would phone and arrange for teachers, chapel speakers and student assignment church visits. We ended our time of ministry there with the Bible School at a near-record attendance.

Whenever you have a group of youth living in dorms together, you need some rules. And some independent youth is always "testing" these rules. One young fellow never wanted to do his compulsory work detail. I took him into the office, but surprised myself by saying, "You're going to be in CHARGE of the work detail!" What was I saying? But it worked. He was a natural leader, and now understood much better the trials of leadership.

In our Bible School days years ago, we had "individual prayer times" as well as compulsory chapel. We were required to fill out weekly reports on our devotional time. I didn't like doing that, but my brother David advised, "Just do it. Those who complained last year got nowhere!" So, I complied and challenged myself to a more disciplined and longer prayer life. We instituted two new changes in the schedule: compulsory devotions every morning for at least 30 minutes after breakfast and, on Wednesday nights, an hour of continuous prayer.

Within two weeks our discipline problems ended.

We had been in the Canary Islands over two years, when we got some very welcome helpers. Our son Stan and his family had been ministering in the mainly native Indian island town of Alert Bay, just east of northern Vancouver island. Stan had learned "Kwak'wala", the native language, and worked on his master's degree in linguistics in the University of N. Dakota during the summers, when people left Alert Bay for commercial fishing. He taught Missions to our students, giving them an unforgettable missions experience with a trip to Morroco. Many people from this Muslim country were living in the Canary Islands where they could be freely evangelized. In fact, Stan saw them looking at our Christian TV program when he visited their homes.

Nearby Mauritania, south of Morocco, seemed very closed to the gospel. One lady missionary dared to enter, saying "I don't expect to sow, nor do I expect to reap. I just want to plow." I consider these pioneers to be "unsung heroes." But God had a way of entry to a country with no known Christians. World Vision gained entry by its medical and childcare program. Their Mauritanian secretary was saved, and stayed at our Bible School with her husband. A newly-converted Canarian Doctor and his Finnish wife became our friends. He joined a Spanish governmental aid program to Mauritania, ministering Christ in that way. Recently I heard of some water baptisms taking place. It is refreshing to see our mission fields becoming "missionary-sending" countries. They can live cheaply and there is less suspicion aroused by Spanish or other non-American faces.

The Korean church was another strong missionary body, with several churches on our island. How they got their first large church is wonderful story.

Mr. Chan's wife was a Christian, and kept inviting him to join her in some home meetings held by a Korean pastor. The Koreans have a fairly large fishing fleet based in the Canary Islands and several thousand live on these islands. The Korean pastor insisted Chan come to the meetings and befriended him till he came to Jesus. Now they needed a church building for the growing congregation. Mr. Chan bought a very large fishing trawler and vowed, "I will dedicate the first catch of fish to help buy a church!" The first catch was ENORMOUS, netting over $200,000.

Meanwhile, a large nightclub overlooking La Palma city went on sale. Most of the tourist business had relocated to the sunnier south of the island after a four lane highway was built. The price kept dropping till the Koreans, with Mr. Chan's "miracle catch of fishes", could buy it. Yonghi Cho presided at the dedication. I've preached there several times, with our Bible School students on assignment. This church has sponsored several new churches, helping our Bible School graduates.

Every time I spoke there in the morning service, I'd be invited to a meal, in the basement of what had been a nightclub cafe. I had a hard time liking the "Kim Chi" or whatever their staple spiced cabbage was called. Whenever I'd put it to the side of my plate, a Korean across the table would reach over and spear it with his fork. (Korean table manners, and perfectly acceptable!) I had to preach wearing rice fabric flip-flops, not shoes. I was tempted to lift my feet and show them off to the students, but they'd have laughed uproariously. And we'd never have been invited back. All the congregants had the same large Bible with hymns in the back. Praise to God was loud and hearty, but at the sound of a little bell on the pulpit it stopped suddenly. May God bless the Korean missionary church.

One day we got a call from Jim Nealy, our area director, asking if we could PLEASE use a great team of builders from the Chicago area. Another destination had fallen through. I immediately thought of pastor Santiago Santiago Santiago on the island of Lanzarote. You may wonder about this name. This is how Spanish names work. Say, for example, the father's name is Jose Santiago and he marries a Maria Santiago. Their child's FIRST name is chosen by them. His SECOND name is the surname of his father. His THIRD name is the mother's surname. Thus, Santiago Santiago Santiago, who married a Gonzalez. The children drop the father's third name (from his mother), and would be called for example Jose Santiago Gonzalez. Understand? I get confused too. Father, mother and children all have different last names.

We had ministered a year earlier in Lanzarote in a tiny church that British tourists and expatriates had helped finance. God moved in and several were filled with the Spirit. I suggested he accept this work team. We sowed a little money and the church doubled in space and numbers. A while later, we noticed the place was full but there was a vacant lot beside it. Again we sowed part of the money; they bought the lot and doubled the church again. The last time I heard, they had grown to over 400, (a huge number for Spain), and had started several more churches. Many of the attenders were recent arrivals from South America. The main church growth in Spain during our time there came from three groups: South Americans, Gypsies and former drug addicts. What they have in common is that they are at the bottom of the social ladder. "The poor heard Him gladly." All we did was sow a little encouragement, money and vision. God did the rest.

Can alcoholics and drug addicts really be changed? I should say so! In fact every Bible School student "dorm monitor" we had was a former drunkard or addict. We waited till they were married and in their second or third year of studies. A little residence was provided on campus and they made splendid supervisors.

But many converts carried the consequences of their past. We visited at the bedside of Montse who was a former drug addict. She had been a vibrant Christian who faithfully attended our church. Against our counsel, she took an ill-advised trip with her husband, an addict, to Peru. There she resumed the drug habit and returned home, repentant and crying, "I don't want to die!" But her face was like a death mask, and she had to crawl to get up into our van for church. She knew heaven was near as we prayed at her bedside. We preached at her funeral.

Later we came to know another Montse (named after the nearby Catholic Shrine of Montserrat). She was a precious sweet and tiny little lady, like a china doll with a pixie face. She never took drugs, but lived with a man who did. Soon she realized she had AIDS and we heard again the plaintive cry, "I don't want to die!" But when these druggies are truly saved they get on fire for God and work to bring their entire families to God in the time they have left.

The Christian drug rehabilitation centres are a strong witness for Christ's compassion. The federal government has a HUGE treatment centre, built like a jail, in Madrid. But the "graduates" go through a revolving door and soon return. At a community gathering I heard a lady complain, "The biggest drug is alcohol and nobody does anything about that! The only ones that are doing a job, and doing it for free are the Evangelicos." She said it very reluctantly, but they all knew it was the truth.

Another ministry we took over was meetings with the many tourists who visited. A philanthropist had built an ecumenical chapel in Las Palomas, near the sand dunes in the south. Different groups took turns providing Sunday services at set hours. The Lord saved some locals working in the hotels, and we began a mid-week Bible study for them in a hotel lobby. We were able to get a retired British minister to take over this ministry on a year-round basis.

One day, shortly after my arrival on the islands, I found an in-depth study of the island's growth centres. It was meant for investors, but I got excited as I saw the possibilities for the gospel. We shared this information with the other pastors. Finally, I challenged my students to hold a campaign in fast-growing Mas Palomas, a city of nearly 30,000. Jorge Bardey, from Argentina, once again gave us a good campaign. Our little church that used to meet in the Bible School chapel had moved to the nearby town of Santa Brigida It also doubled in size when, under Bardey's ministry, God healed several folks who then told others.

A recent graduate took over the church in Mas Palomas that sprang from that campaign. We rented a space in a shopping center, not far from places that offered body-piercing and other exotic services. People would wander in and listen out of curiosity. I won't soon forget the man who sat, transfixed and said as he left, "God the Father and His Son were here today!" The island, like many tourist centers, was full of drug dealers and casual users. We had at least ten drug rehabilitation centres on our island of Gran Canaria, with its one million residents.

"Princess Diana died in a car accident!" The news spread almost instantly across the island. In the tourist city of Mas Palomas the British tour-operators pleaded with us to hold a memorial service for Diana. "We'll spread the word in every hotel!" they told us. The next day, the Catholic priest gave his regular Bible reading and homily for the day, completely ignoring Diana's death. I was joined by two ladies, one Irish, the other Scottish, as we presented our tributes to a packed church. This was one of the most difficult and delicate messages I've ever preached. It would not do to besmirch her memory, but I could call on people to prepare for death while they are still living.

Starting new churches ought to always be a part of any serious missionary endeavour. It may be via direct evangelism, or indirect, such as training workers in Bible School, writing gospel tracts, Bible translation, or building. However you do God's work, persecution should be expected. When we were preaching in our first tent in Salta, Argentina, a man came up the aisle throwing a heavy car part at me. But the missile kept falling short, into the sawdust. Ushers finally ushered him out. That same tent was burned down in another campaign in San Juan. Fortunately it had a steel structure that survived and the canvas was replaced with the tent we used in Jujuy.

A national minister friend of mine had a tent burned down while he and his family were guarding it. They barely escaped with their lives. In Mendoza a stone was hurled through a window in my direction. I kept the stone as a souvenir for awhile. In Spain persecution took the form of exasperatingly slow approval or the downright refusal to permit public meetings. One pastor on the Canarian island of Fuerteventura threatened to protest by printing leaflets comparing local leaders to the inquisition. The leader "caved" and now this church is one of the largest in Spain.

God has his own ways of dealing with spiritual opposition. In Barcelona province, a Christian radio station kept getting shut down. The bold evangelist did not touch the taped-off transmitters. He just bought a new transmitter and kept broadcasting. The opposing governor's wife suddenly lost her voice and feared it was God's punishment for shutting the voice of a godly radio station. The station is still broadcasting today. Pastor Sergio Zubillaga started a Christian TV station in Gran Canaria which had no end of problems keeping on the air. A government monopoly in the communication field can tie up broadcast permits. They attempted to shut it down because staff was not paying income tax on their "earnings." They simply could not conceive of people volunteering their time for the Lord's work.

Our Argentine evangelist friend, Jorge Bardey, had his heart set on starting a church in Olavaria, a city about 200 miles southwest of Buenos Aires. He'd been an evangelist's assistant some years earlier, and their campaign was abruptly closed down by the opposing chief of police. "One day I'll return!" he vowed.

He made careful and prayerful preparations, including getting a federally-issued church-meeting permit. As the campaign progressed, news of remarkable salvations and healings reached the ears of the police chief. He sent two of his henchmen to say "In two days, we're shutting down this campaign!" Jorge, prayed up and buoyed up with spiritual empowerment went next to the police station and declared, "I want to talk to the chief of police!"

"He's sleeping!" they shot back.

"Wake him up!" Jorge demanded.

Now it must be noted that "the siesta is sacred" in Argentina. You don't interrupt a siesta. The police chief, a large man, came groggily and ill-humoured to the desk. "What do you want?"

"Sir" Jorge said, "It has come to my attention that you want to shut down our evangelistic campaign. I have here in my hands an authority from the federal government to hold services. If you so much as touch us, I'll report you to the government for abuse of authority! Good day sir!"

In under two weeks the offending chief was transferred out. And the church now has nearly 2000 congregants and 16 branch churches. This is "Holy Ghost boldness" like the apostles prayed for and God's answer was to refill them with the Spirit (Acts 4).

Joseph G. Anonby

Ministry Meditation:

The much-persecuted Apostle Paul wrote "... all that will live GODLY in Christ Jesus shall suffer PERSECUTION" (2 Timothy 3:12). This is especially true for those who are on the "front lines" invading Satan's territory. We read in Ephesians six about the complete armour we need to fight the good fight. Of course we need to recognize, as in the examples above, that opposition takes many forms.

If you have no struggles at all, check to see if you're just "coasting." Those who live seriously godly lives have a social conscience. Wilberforce with Wesley's help got slavery outlawed. Elizabeth Fry reformed terrible prisons. Many of the early labour unions, which still have "Christian" in their names, fought for fair wages and working conditions. Rev. Martin Luther King fought for racial equality, knowing his life was in danger.

In the beatitudes, Jesus promises special rewards: "blessed are they which are persecuted for righteousness' sake...rejoice and be exceedingly glad, for GREAT IS YOUR REWARD in heaven, for so persecuted they the prophets" (Matthew 5:10-12). You are in good company if you lose the "popularity contest" in school or the workplace. It means your spiritual stance has been noticed.

Meanwhile, all the other missionaries in the Canary Islands had either retired or were transferred elsewhere. While on a "mini-furlough" due to having nobody to replace me as Director, the national church leadership went to Columbia. I mean ALL the leadership, (about 58 people), from an organization with only 15 churches. They got very enamoured about Cezar Castellano's "Cell Church" vision in Bogota. This was now about the sixth overseas pilgrimage made by our leaders looking for the "magic bullet" to grow churches.

The more I read and listened to the promoters, the more "red flags" were raised in my mind. It was partly a "rehash" of all the "new wave/superior truth" divisions I'd experienced over the years, starting with our first church in Gibsons. They professed to be against denominations, but were well on the way to forming their own, with the "Group of Twelve" leaders spanning the globe. It used business tactics similar to "company takeovers." There were special financial incentives for leadership, "shepherding" (dictatorial leadership), prosperity doctrine and the obligation to order many supplies from Bogota. It looked like they'd assume our church leadership if they could. The superintendent asked for them to come to our national conference which was delayed to accommodate them. But the Colombian leaders demanded our denomination bring both husband and wife, and house them in a five star hotel!

We had begun two churches with our student graduates. In six years, one half of all the pastors were students who had sat under our teaching. There were some tense moments as I attempted to debate the merits of adopting a new system. Bible Schools, the bastion of doctrinal teaching, were to be attended only by students who had taken a prescribed course on cell church ministry. They could not advance to a second year unless they started more cells. Retreats were held and the content of the teaching was not revealed. You had to pay for and personally attend a retreat to find out.

There were positive aspects in the discipling of new converts. But all the regular departments of the church such as Sunday School, youth and women's minstries were replaced by cells. Students who had graduated from Bible School were given limited places of ministry till they took the "cell course." Would-be pastors were told not to expect a church, but rather to be "good seconds," while a very few lead pastors assumed control of several weaker churches in the name of "helping" them. The business world would term that "a hostile takeover."

In short order, nearly all the non Assemblies of God cooperating churches pulled their students out of Bible School. I was asked to resign as the Director, but to stay on and be "an apostle" since I'd started churches. We went to prayer about it. I was inwardly seething at the injustices and abuse of leadership authority. After consulting leadership we trusted, we decided to move to another field till things calmed down.

My heart was very grieved with all these activities that I believed were harmful to healthy church growth and governance. I was driving to our newly-founded Mas Palomas church and got into a traffic incident. I was already upset, and this made me very angry. Suddenly I felt a sharp pain in my throat and a general nausea. Fernando, our student pastor, phoned ahead and drove me to the hospital. I chewed some aspirin en route, since I believed it was a heart attack. It turned out to be an acute angina attack and I was treated and hospitalized. It was a "near thing" to a heart attack.

After failing a treadmill test for the second time, I was scheduled for a double bypass operation. But there was "no blood" in the hospital for the operation. Leadership that had wanted me ousted now made a special appeal to all our churches, pleading with people to give me blood. Many people gave blood which was a major testimony to health officials. They had presumed we were like Jehovah's witnesses, who due to a biblical misinterpretation refuse to accept blood transfusions. It moved my heart to see how God arranged reconciliation in a most trying circumstance. God was healing my broken heart.

Finger of God at Canary Islands

Volcanic rock formation

Bible students in Grand Canaria

Bible school graduation

Chapter 13

To the Dominican Republic (2003-present)

Now the question was "What field do we go to?" To my surprise, the leadership in Argentina did not see a need for me to return there. We were asked to be Bible School directors in Jamaica, but preferred a Spanish-speaking country.

Caribbean Field Director Dick Nicholson, whom we 'd known from Argentine days, asked me to consider the Dominican Republic. "They'll love you there!" he said. Well, that would be a treat and a contrast to some of the argumentative Spanish leaders we'd known. They were looking for missionaries who were experienced in Bible School work. After a period of convalescence and itineration, in January 2003 we were bound for Santo Domingo, capital city of the Dominican Republic.

Haiti is on the western side of this Caribbean island called "Hispaniola." The Dominican Republic is just south of Cuba and Puerto Rico. It is 50% larger than Vancouver Island and about one-half the area of Pennsylvania. In 2013, the population was about ten million people including over one million (many undocumented) Haitians. Haiti has close to nine million people, many in abject poverty both before and after the 2011 earthquake.

As in the Canary Islands, tourism is the lead industry, with about four million visitors, especially from Canada and USA. Over one million Dominicans live on the eastern seaboard and there is a constant flow of visiting relatives forth and back. Close to 30% of the Dominican economy depends on "remittances" from relatives living in New York. (New York is so important that a young boy once asked me "Are you from New York or the United States?") Clothing is manufactured in numerous "free trade zones." Another important industry is agriculture, (cocoa/chocolate, avocados, sugar, tobacco, coconuts, coffee). While driving on back roads we often see cocoa laid out on the asphalt road to dry before shipping it to market. Mining for nickel and gold is a billion dollar resource industry despite environmental protests.

We began attending a rented small local church under the ministry of Agapito Franco. It had a leaky tin roof and very limited space. One missionary despaired of the church ever growing to over 40. But the Lord laid on our hearts the goal of of building them a church. Agapito was overwhelmed and anxious to "seal the deal." "Did you hear that?" he shouted, "Did you hear that? The Anonbys said they are going to build us a church!"

We weren't certain ourselves how it would be done. The area was close to the site of the

Pan-American Games. The Dominican heroine was Yudelka, a champion lady weightlifter. Sanchez, a gold medallist in the hurdles became their hero. The names of their Big League baseball heroes are household words. One was trained by his Christian mother till he made the big Leagues. Children use a broom stick to hit water-bottle caps in the narrow streets. After that, hitting a baseball is easy!

Agapito found a 1000 square foot house for sale. It was occupied by four families who were all renting, but nobody paid the rent. The disgusted owner put it up for sale,with one deadbeat renter still in residence. We refused to buy until the place was empty, due to sad stories we had heard over the years. The reluctant landlord PAID THE TENANT TO LEAVE, and the house was ours to become a new church.

Pastor Stepp from Bethel Park, PA brought a group of 23, including a spry 83 year-old lady. Pastor Dan Eagle from New Westminster, and childhood friend Ken Hood of Castlegar, B.C. gave both financial and work support. The Bob/Betty Sahlstrom and Paul/June Kavaloff families (from Castlegar), brought further friendship and ministry. Nobody can yodel like Paul! One or two experienced national builders, plus many volunteers were eager to help. Someone usually asks, "Wouldn't it be cheaper to just have the church send money, and put the airfare into buying materials?"

But what happens with a work/ministry team is more than just a transfer of money. Youth feel the joy of serving and are called to missionary work. Nationals form friendships with smiles and simple acts of kindness. Today, youth the world over are linked together with "Facebook" and other social media. The nationals are proud to have Americans and Canadians as their friends. Prejudices begin to crumble.

Older people, many of whom no longer have children to care for, may adopt or sponsor an orphan or student. When they see the difficulties a typical missionary faces, they return home and give more generously to missions endeavours. They also have a better understanding of what to pray for.

For example, Bibles are an expensive purchase for someone with hardly enough to eat. Programs such as "Light for the Lost", "Speed the Light" and "Wing the Word" support the import and purchase of vehicles and Bibles. Several groups have taken a suitcase each of pastoral Study Bibles which are provided to us at wholesale prices. We've become the "Bible people" to get these "New Life Study Bibles" to each of our students. We also work with the Bible League to get inexpensive Bibles into the hands of new converts. Every time I go to Bible School I'm asked for Bibles. People pay a token amount which we reinvest in more Bibles and literature.

"Sunday School is the heart of the church." That's the oft-repeated theme we hear every conference. The Sunday School Annual Convention is a much-attended event. We often house visiting speakers, who marvel at the large Sunday Schools. Neighbourhood children are evangelized, and then bring their parents to church. I was astonished at how much the church youth know about the Bible. Some did better than I on little details. And the reason for that is constant Bible study in Sunday School. This is a wonderful cure for "biblical illiteracy" and the foundation for an adult moral framework. All of it is woven together from an early age in an impressionable child's mind.

As a child is developing, his parents are usually the chief moulders of his character. Children are taught by example, story-telling, Bible-reading in family devotions and regular church attendance. Parents who are inconsistent in their "spiritual exercises" also teach a negative lesson; God is not the priority of their life.

We read that "Jesus increased in wisdom and stature, and in favour with God and man" (Luke 2:52). His cousin John "grew and waxed strong in spirit" (Luke 1:80). Jesus, like John, spent some quality time with God "in the desert." Jesus "returned in the power of the Spirit" and began his public ministry saying, "The Spirit of the Lord is upon me, because he hath ANOINTED ME TO PREACH...TO HEAL...TO SET AT LIBERTY" (Luke 4: 14, 18).

Ministry Meditation:

What is "the anointing" of God's Spirit? How do you get it? What does it look like? There are a lot of questions which aren't always easy to answer precisely. But, as I mentioned earlier, when the anointing is present in "power evangelism" people take notice. I'm still learning, at age 70+, how to live under the anointing of the Spirit. But I'd like to share some of what I have learned.

First, we need to humbly recognize that God is sovereign. "God set the members every one of them in the body, as it hath pleased him" (1 Corinthians 12:18). Prophets such as Jeremiah and John the Baptist were foreordained to their ministry before birth. Others, like Samuel, were dedicated by their parents to God's service. God chose to accept and anoint their child.

Early childhood experiences can foster an anointed life. Samson took the Nazerite vow, which governed what he ate, what he touched and even how he wore his hair. Samuel lived in the temple as the servant of Eli, the high priest. He probably didn't play a lot with the neighbourhood children.

John Kilpatrick, was a young teenager when God spoke to him as he watched a biology film in school. Time stood still while God told him of wonderful plans He had for his life. But he was cautioned, "You must pick your friends carefully, or none of this will happen!" Simultaneously his grandmother had a visit from a lady who sold home-baking. She said, in essence, "Your grandson John has a special mission from God, but tell him to watch the company he keeps, or it will not happen." When John returned from school, they shared and confirmed their matching visions. John went on to be used mightily of God in the Pensacola, Florida revival that had world-wide influence.

I'd like to suggest we look at two main challenges in seeking the anointing of the Spirit. I mentioned this previously, but it is so important that I will be making additional suggestions now.

I. HOW TO RECEIVE THE ANOINTING?

"Thou hast asked an hard thing" Elijah warned Elisha when he plead, "let a double portion of thy spirit be upon me" (2 Kings 2: 9,10). Here are some practical ways to improve the quality and frequency of God's touch.

1. *BE BAPTIZED IN THE SPIRIT. It is a biblical command to "Be filled with the Spirit" (Ephesians 5:18). The purpose is clear; "Ye shall receive POWER after that the Holy Ghost is come upon you and you shall be WITNESSES" (Acts 1:8). This is not to say that God can't do a special miracle. Our Salvation Army widow neighbour, Mrs. Brown, told how God gave her the Ukranian language for about 20 minutes to explain to a dying friend how to get saved.*

2. *DESIRE THE GIFTS. Paul exhorts us to "Desire spiritual gifts...to EDIFY the church" (1 Corinthians 14:1, 4). Note the motivation is to fortify the church, not ones own reputation.*

3. *CLEAN YOUR MIND. David said "Search me O God, and know my heart: try me and know my thoughts: and see if there be any wicked way in me, and lead me in the way everlasting" (Psalm 139:23,24). Surround yourself with God's Word by reading the Bible, listening to Christian music, reading wholesome books and having Christian art or posters on display.*

4. *WORK FOR GOD. There's no point in filling a car's tank with gas if you're not going anywhere. When we are busy for Jesus, we can call upon Him for his enabling power.*

5. *STAY IN CHURCH. Isaiah was in the temple when God called "Whom shall I send?" Samuel, Joshua, Simeon, Ana and many others were "where the action was." Jesus lashed out at priestly hypocrisy, but still attended the synagogue "as his custom was." The true church is God's body and Jesus is the Head. Loners are usually ineffective over the long term*

6. *CONCENTRATE on being Christ-like and winning souls. That's the "Great Commission" for the church and the Christian's purpose in life. It's the reason God gives "anointing that breaks the*

yoke" to break down Satan's rule and establish God's kingdom.

II HOW TO MAINTAIN THE ANOINTING?

7. *CONTINUE in the steps you followed that brought God's touch.*

8. *AVOID temptations that fritter away God's power. This may include people, places and activities that consume your time and interests.*

9. *DON'T "FAKE IT". The anointing oil had a unique composition that was not be be duplicated for other uses. Sometimes I cringe when people shout, shake or imitate the anointing, yet in my spirit I sense it's phony.*

10. *DON'T ABUSE IT by flaunting it, using it for "money for prophecies," or other personal benefits.*

11. *THE LETTER KILLS, but the Spirit gives life. NOT being empowered by God makes Christianity unattractive and boring.*

12. *THE DEGREE OF ANOINTING can vary, depending on the atmosphere. "He could do no mighty works there because of their unbelief" (Mark 6:5,6). Little prayer=little anointing. Get allies to pray with you.*

> **13. IT'S A MIRACLE EVERY TIME.** Never cease to seek God's anointing and never stop thanking God for using you.

Some final thoughts about anointing. The "Anti-Christ ('anti-anointed one')" hates the power of anointing that puts the fear of God into demons who tremble. I noted in Argentina, that when I was speaking under a powerful anointing, demons would begin to manifest. (People shaking, shouting, disturbing the meeting and bringing attention to themselves.) Satan tries to duplicate the anointing. The hair on the back of my neck would stand up as I listened to radio tapes of Hitler and Eva Peron speaking to thousands. Hitler was in process of being presented as a messiah to Germany. Eva Peron was a spiritist, and her successor, Isabella Peron publicly invoked her departed spirit as crowds chanted "Eva, Eva!"

God ordained a special anointing upon priests, kings and prophets being inducted into ministry. Saul, was met by a company of prophets "and the Spirit of God came upon him, and he prophesied among them" (1 Samuel 10:10). But pride, disobedience and jealousy ruined his relationship with God. An evil spirit possessed him, and one of his last acts was consulting a witch. Balaam was a covetous prophet who sold out to the highest bidder. He was killed in battle fighting against Israel. In prophetic irony, Micaiah sees, "The Lord hath put a lying spirit in the mouth of all these prophets"

(1 Kings 22:23). The wicked king Ahab, convinced of victory by false prophets, went to his death in battle, despite disguising the fact that he was a king. Anointing is wonderful, but power tools need to be used with care.

Returning to our move to the Dominican Republic, we decided this time to take only 10 suitcases. Years ago, when we went to Argentina, missionaries were allowed three 200 litre barrels for each adult and one per child. We brought twelve barrels, and personally paid for the extra three. Senior missionaries advised us of the poor quality of clothes manufactured there. For many years, those barrels did service coming and going to the mission field. We still keep a few for storage, lined with large plastic bags in our garage. Some of our heavy winter clothes were still intact after 13 years in these sealed containers! Today, many find it cheaper to import all their household furniture in containers. Later, when they leave, they are sometimes able to sell at a profit. We were usually able to buy good used furniture from people moving out. And when we returned the last time to Canada we were able to furnish much of our house for under $1000 from several nearby thrift shops.

Ministry meditation:

Nationals are very observant about the size and furnishings of a missionary's home. I was once asked, "What is the size of your home in Canada?" When I showed him a picture, and told him renters were helping to pay the mortgage, he was quiet. It seemed he'd thought we were "living large" overseas.

Our home overseas, especially in larger cities, sometimes came already furnished. If not, we often bought used furniture from missionaries leaving. The Pentecostal Assemblies of Canada would pay for a fridge and stove and little else. The Assemblies of God paid about $8000 per couple in the years 2000-2015. If you paid less for shipping and buying, the balance of money could be saved. In the Barcelona area of Spain, the "Catalans" put used furniture on the street, lest people think they were so poor they had to sell it. I picked up a lot of that "free stuff!"

I have heard nationals complaining of missionaries who have opulent furnishings. Especially if they were inhospitable. We had large amounts of people visiting who were new converts; sometimes poor and shabby. They felt comfortable in our house, and we had few thefts or furniture damage. Anyways, it was cheap, and we'd likely be moving. For many years we never stayed in a place over three years. It reminded me of an old army captain who told his soldiers, "Don't put your tent stakes down too far, we'll be moving on in the morning!" The world is not our permanent home. Be careful of where you are putting your treasures. That's where your heart is!

The principal missionary families were the Coads, DeFreitas, Martinez, and Jamie Bello. The Coads, very kindly allowed us to stay in their home while we looked for a house. It was the first time the AG mission had bought a home in the Dominican Republic. Nationals wondered why we had not done so previously, but funds were not always available. The house was situated in north Santo Domingo, near a botanical garden and an English school where many missionary children had attended.

When you buy (or rent) a house in the mission field it often comes as "just bare walls." Can you imagine a house with no stove, fridge, cupboards, light fixtures, hot water, window bars, or any kind of light fixtures? Just bare wires hanging from the ceiling! It took a lot of work to make the house liveable, and satisfactory for other missionaries. We also had to install an "inverter"– an electrical backup device with four expensive ($1000) batteries. This is a "must" in the Dominican Republic. On my last visit we had no electricity for nine hours each day. Some power outages lasted 36 hours... not very easy in hot summers, with no working fans after the battery power ran out.

We got a Mazda 4 wheel drive vehicle, quite a bit smaller than expected, but the PA district youth department was low on funds. We installed a small backward-facing movable "trundle seat" in the rear hatchback area. Youth would fight to sit there and dangle their feet out the rear window! There are very few "safety checks" in the Dominican Republic

We visited and ministered in several churches, some numbering in the hundreds. Calvary Church, a large church on the main street ("Lincoln"), began in missionary Lesterjet's garage where his daughter started a Sunday School. The Bible School was built in the 1940's by Peterson, who bought a large tract of land out in the country. In due time it became a very valuable property situated on the main highway. I was asked by Dario Mateo, the director, to teach in this residential school with about 80 students. I was also asked to direct and teach in a night school about 50 kilometres from Sto. Domingo. I turned down that offer. We were both accustomed to "live-in" schools and thought they were the most effective for preparing ministers.

Part of the job of being a teacher included driving students to visit churches and promote Bible School. We made some very long and eventful trips. On one trip I got up at 4:30 a.m. and drove an 18 passenger van that dropped students off at eight different churches. We stopped for lunch at a place that provided a meal of goat meat. I chewed on a piece of jaw with the teeth still intact!

Our final stop was at Jemini, near the Haitian border. This place was HOT and also was in the middle of a mosquito epidemic. Local stores had run out of repellent. What to do? In church, the folk stripped branches off some trees and swatted themselves. I wore a long sweater and a lady behind me swatted my neck. But when it came my turn to preach, I got eight bites on my neck! I decided it wouldn't be the best time for an altar call. But director, Dario Mateo, took over and called people to the altar. Nobody came! We both stayed the night in a small local hotel, and were relieved to have a working ceiling fan that drove away mosquitoes. Dario was allergic to bites, and got sick the next day. I dubbed the town of suffering "Gethsemani" instead of Jemeni!

As non-natives we are more vulnerable to local diseases. On two occasions my legs became red, hot and swollen. I was sick and feverish. The doctor said "It's cellulitis!" What? That's for plump ladies! Well, that's what they called it there. The doctor's advice was "Don't walk around in bare feet; mostly it's foreigners who get this ailment, and it can be very serious!"

We always advise our visiting teams, "Don't drink the water!" A father and son had the habit of drinking the water as they showered. Both got VERY sick, and the father was wearing adult diapers until we got him to the hospital. He got partially cured, but did not continue using the medicine once he arrived home in Canada. Big mistake; he got sick all over again for several days.

Avoiding dysentery can get complex. Ellen learned to methodically soak lettuce and other fruits or vegetables in purified water with iodine for 20 minutes. Then rinse them in water that is also purified. Even watermelons need to be washed, or the knife that cuts into it will bring surface dust into the fruit. Some missionaries leave out those steps and lose days in the hospital. Malaria and newly-discovered insect-borne diseases like the Zika virus also occur. Those tropical bugs are fierce!

Missionary Kirby Jennings, who spent some time in west Africa, told of a man who gave the following testimony: "I had very bad constipation. But one day a leopard jumped into my path, and I was instantly cured!"

One keeps meeting up with unexpected creatures, great and small on different mission fields. In Buenos Aires, Argentina, we had seven different colonies of ants in our house. We had to clean the counter with vinegar several times a day. In the Dominican Republic, the most "tropical" of our postings, we had flying ants after big rainstorms. They'd leave their underground nests and fly to wherever there were lights. We'd have to curtain off every window and dim our lights. Then they'd crawl under the sliding windows.

I also had some alarming sessions with "fire ants." When they bite, you REALLY feel it! There was a colony in our yard that laid claim to the front sidewalk border. I gave up on putting in a flower hedge and planted some plastic flowers for awhile! Sweet revenge on the ants, till Ellen tired of seeing a plastic hedge. Then there were the 1½ inch caterpillar-type worms which marched up to eight at a time across our living room floor. They were so silent that I didn't notice them until they began crawling up my neck while I was reading! In the Spring we'd have about one large (palm size) tarantula each week inside the house. I kept a very large fly-swatter on hand for those encounters. Splat!

Ministry Meditation:

Bugs and missionary health – how do we prevent serious sicknesses? I remember a doctor's lecture at my first School of Missions in Springfield, MO. He gave a long list of ways in which one could catch malaria and other possibly fatal diseases. Even washing dishes had to include a special drying process to prevent dangerous diseases. The class became silent. I could feel a palpable fear gripping the candidates. At the end of the lecture the Lord gave me a public prophetic word. The Lord dispersed the fear with words that said, in essence: "You are going out in my work, and I will be with you to protect you from danger. Do not be fearful. God is with you and you are under my special care."

The safest place to be is in God's will. Buzz Aldrin, one of the early astronauts, went safely on a dangerous mission to the moon and back. But he fell in the bathtub back on earth and fractured his skull! "Fear not!" is one of the most repeated phrases of the Bible. I guess we all need reminding!

One of my youthful interests was learning about ways to save time, (so I could do "fun things" like reading an entire book in an evening). I bought a book on "high speed math." I also took a course on "speed reading," shooting for 1000 words per minute. So the Lord must have a sense of humour in letting me live in "mañana" countries. Can you imagine five hour services, called "Jubilocs," the last Sunday of every month? You are also expected to fast till the meeting ends. Then there are the four hour waits at government offices and the week-long wait for many car repairs. (Or the 4 month wait for insurance repairs in the Canary Islands after our parked car suffered a "hit-and-run".) I could add a lot to this list, but you get the idea. I've learned to use these times to pray, or read. I always keep a book or magazine in my car.

My daughter Joy and husband Ken were visiting and locked the keys in their car. They asked a tourist security policeman, "What do we do?" He found a well-known car thief nearby who opened it quickly, under his watchful eye. Of course they had to give a tip to the thief, and, probably, the policeman. A visiting team from PA used a missionary car that locked automatically...with the keys inside. A former (now converted) car thief showed us how to enter with ease. Even thieves can be useful!

"The library is the heart of a college." This is a common saying in educational circles. As I continued working with our evening or Saturday Bible schools and attending their graduations, I noted the critical lack of study books. The Assemblies of God had a missionary, Rev. Everett Ward, who specialized in getting libraries set up in overseas Bible Schools. I took this on as a project, with many churches and individuals helping us by donating "mini-libraries" for about eight branch schools. I remember telling students, "One day you'll be mostly reading from internet books." They looked at me in disbelief; only one student had a computer.

But in a few years, how that has changed! Now my challenge as a teacher is getting students to "footnote" their literary sources, most of which were "online" references.

Many of our students had hair-raising experiences with criminals. Newton, one of the first batch of six students, was returning with friends from an early morning prayer meeting. The group was cut off by a motorcyclist, who waited for his leader to take their wallets. The leader asked, "Are you evangelicos?" When they said, "Yes!" he let them go, saying, "I don't rob evangelicos!" Another told of a church member who visited his bedroom at night and tried to steal 6000 pesos (nearly a month's wages at the time) from his wallet.

I was walking near the marketplace when a mumbling man approached me. He grabbed my arm and dug his fingers hard into my flesh. He left five bleeding gashes. I believe he was demon-possessed. Onlookers told me, "Yes, he's bothered a lot of people." Over the years we've prayed for a number of demon-possessed people. Some have "inherited" fortune-telling and other such "skills" from their grandmothers. We find it best to pray for those afflicted in a church setting, with mature people of faith giving support. Usually it is best to remove them from the sanctuary, because Satan likes getting the attention by removing the focus of salvation preaching. Idle onlookers at times become fearful, and stop going to revival meetings.

Satan has many strategies to bring defeat to churches. One that I've observed many times is divisive doctrines. I spoke to several main leaders about our Canary Island experience with the G-12 "Cell church"; a concept of "the 12", like Jesus' disciples. Ceasar Castillano, of Bogota, Columbia, had a vision of how to multiply these cell groups into a large church. These leaders said, in essence "We won't have problems here about that; we preserve sound doctrine!"

But after a few months passed, I was dismayed to see several leaders attempting to install what I had just warned them about. I spoke to a teachers' meeting about our Canary Island experience, and challenged them to "grow up and find a Dominican method of growth." In a few hours three people informed the Superintendent of my speech. Our field director was told that "missionaries should be quiet."

The denomination's unity was soon at risk, with people vying for their G-12 doctrine. Secret meetings with secret teachings were held, and pastors often got big offerings as part of the program. A group of four previous superintendents published a "counter-teaching" circular that documented several unscriptural excesses. G-12 sympathizers tried to interfere with elections. Anyone who was not "on-board" with G-12 was not invited to teach. Recent graduates no longer were given churches to pastor. It took several years to settle down, and I had to "keep my head low" in the process. One of the proverbial "seven deadly sins" is "he that soweth discord among brethren" (Proverbs 6:19). It's a lesson that we have to keep learning!

Most of the regional districts have monthly pastors' rallies to pray, hear testimonies, listen to

Anonby Adventures in Missions

teaching, and encourage each other. I attended whenever possible, and gave a teaching about financing the local church. We heard a dramatic testimony of healing from a man called "Castro." This is his story: "Someone gave me a prophetic word; 'You will go through a process of suffering'." Soon afterwards, while the family was crossing the "Malecon" (a busy sea-side street), they were hit by a speeding car. Castro heard a bystander say "The girl is dead." Immediately he rebuked death and she revived.

Castro was taken to the hospital, where they found no broken bones, but very damaged lungs. At first he was swollen up, but as weeks passed, he dropped to 85 pounds. One doctor said, "Turn off the machine!" Another responded, "We've already spent a lot of money on him, but he's an 'evangelico'; he believes in God, and talks about miracles!" Soon after, Castro had a vision of the desert turning green, and heard a voice saying, "The process is over!"

He yanked out the tubes, and jumped out of bed. At the same time, a heart patient in the room was healed and gasped, "He's walking, he's walking!" Castro was examined by 26 doctors and told "You're going home!" He still had an unpaid bill of U$ 3600 which was marked "Paid in full" by the presiding doctor.

Ministry Meditation:

Personal testimonies are a powerful tool for evangelism. Paul often gave his personal

testimony when he began churches, spoke to the courts and to kings. The Bible says "Let the redeemed of the Lord say so" (Psalm 107:2). We miss a blessing from God if we are silent. And needy people miss a word of hope or a lesson about the abundant Christian life. I grew up in a church where we always had testimonies as part of the church service. Most churches did the same. We also had "scripture showers" in which people recited a special verse, and sometimes gave a comment. This added congregational participation to a service. I miss those testimony services, in which ministry came from both the platform and the pew.

On Mondays we often took a day off, and sometimes did "off-roading" with our 4 wheel drive missions car. We'd follow our somewhat unreliable maps in loops through the rugged countryside. Once I chanced upon a tree that had WATERMELONS hanging from the branches! It was called a "cow-elito" tree, but I don't think the fruit was as heavy as watermelons. We loved to eat mangoes, which are orange, with a sizeable pit inside. Some varieties can be the size of a cantaloupe and have both male or female fruits. (The male fruit is more hairy/stringy and sticks between your teeth.) Imagine our astonishment at seeing children doing target practice, throwing the super-abundant fruit in the country lanes. We often followed trucks loaded with impossibly large numbers of country people going to town.

In the remote areas you could find some most unusual items. Can you imagine a radio that runs on kerosene? Or a typewriter with only ONE KEY, that you swivel to tap out a single letter at at time? I got that typewriter at an auction. We would sometimes spot a little evangelical church, (usually Pentecostal), and drop off some Bibles we carried. Frequently roads were damaged by rains and bridges washed out. We would put the car into 4 wheel drive and make a dash for it through water that was close to waist-high. Amazingly, we never got stuck in the Dominican Republic...but I think we damaged the motor mounts once. The car sounded like a tractor!

Most missionaries have dramatic "travel stories". Our plane caught on fire on a trip to Argentina, but we got off safely. A missionary friend from England was so sick from dysentery that the villagers took him to a doctor on an oxcart. The Wellborns, from El Salvador, stayed with us during a "Kings Castle" conference. First they lost their tickets and passports, which delayed their return. Then they were placed on a LONG standby in Panama. Lexus, their four year old said "Mommy give me the tickets!" He placed them in his hands and prayed (LOUDLY), "Dear Jesus, tell your Daddy we want to go home!" The Lord, (and a ticket attendant), heard that prayer and they got on the next plane.

In Africa, the father of Austin Chawner travelled long distances on poor roads in the early 1900's. Once he was so tired that God shortened his journey by a miraculous instantaneous transport home. (See Phillip's experience in Acts 8: 39, 40 "the Spirit of the Lord caught away Phillip...he came to Caesarea"...about 100 miles north). A Liberian missionary ran for many miles at astonishing speeds over a dangerous jungle trail, arriving at a friend's village before nightfall. While piloting a small plane, he heard "pull up, pull up" as he flew through a tropical storm. He promptly obeyed, and just missed a huge power line in the jungle.

In his youth he had twice felt a prompting to witness to someone, but delayed. Both those people died soon after. So he learned to listen to God's "still small voice" and wrote a book with that title. It's important that your best friend is a God who promises, "Lo, I am with you always, even unto the end of the world" (Matthew 24:20).

Missionary Kirby Jennings told us of a similar experience. "While we were missionaries in Mozambique I awoke and felt an urgent need to make a water storage facility at the Bible School compound. I worked long and hard at it till it was completed. Two weeks later, the area water system was sabotaged due to a civil war. The town had no water for three weeks, but we still had plenty at the Bible School!"

Sometimes God's promptings are heard by children. Woody, a construction missionary from Georgia, told this fascinating story of his salvation experience. His eight-year-old girl asked, "Daddy, for my birthday present, could you spend six hours a week with me?" (Who can say "no" to such a sweet request?) "Sure" he replied promptly. "Okay, then come to church with me two hours on Sunday morning, two hours Sunday evening, and two hours on Wednesday!" Trapped! Woody concluded, "In under two years, I got saved."

Ministry Meditation

How can one become better equipped to "tune in" to God's leading? Of course the Bible has the answers to all our spiritual questions. But Paul still found it necessary to write and "stir up your pure minds by way of remembrance" (2 Peter 3:1). In addition to preaching, God sometimes uses incisive prophecies stirring us to "tune in". For example, the Lord gave me a prophetic word at a missionary conference reminding us all to "spend time in prayer, and you will be a vessel open to listen to my guidance. Then you will be able to influence and bless others."

When we were not pioneering a church, we were often invited to churches pastored by our students. Some lived in remote places and difficult situations, such as staying for years in a room of a deacon's house. I would encourage the congregation to donate cement blocks for a parsonage, and I would give matching funds. If the church was near herds of cattle, we often had flies as "visitors." Once Ellen and I left a meeting with 31 itchy bites each from those pesky black flies.

But we were repaid with the blessed presence of the Lord. We would sing, preach, counsel, and pray for people to receive the baptism in the Spirit. The Spirit was often poured out in these services, and seekers beamed as they glorified God in a new heavenly language.

Chapter 14

Beginning the "Deep River Church" in Sto. Domingo

We were asked by the Dominion Republic Assembly of God to come and help them with their Bible School. Earlier missionaries had started many churches, sometimes from humble beginnings, meeting in a garage, in a living room or under a tree. As the years passed, the Bible School graduates became church starters and missionaries were asked NOT to start churches, but instead provide other ministry skills. By this time in our life we had begun 13 churches in three countries. But the new superintendent though removing our main Bible School involvement, asked "Why don't you start a church?" Well...why not?

We began by holding Bible studies in a 'barrio' (neighbourhood), where our part-time maid lived. Damiana had served several missionaries and heard the gospel clearly. After awhile, many of her relatives on the former farm, now a housing barrio, came to faith. Damiana finally made a decision for the Lord. We shifted some of the meetings to our home, and used our own barrio's gazebo for many activities.

The Bible School students got involved in our meetings, and helped in music, drama, games, Christian videos and preaching. We used the "Alpha Program" of basic gospel truth and opened up times of discussion. In the early stages, we ate a full meal every Sunday, saw the video, and then had a dessert after the discussion. As the group grew, we were serving over 300 meals or desserts every month. The folks were very helpful in washing dishes and cleaning up, to make it easier for Ellen.

Then there was a murder. One of Damiana's many brothers got drunk at a party. He and a friend beat up and killed a man. They then fled to the country to escape the "blood avengers". Damiana and her husband, Angel, fled to his house (another former farm that became a barrio). So we began another study group there. We were now transporting several carloads of people to meetings in our home.

Angel's grandmother was the owner of the old farm which was now a barrio. All her relatives lived in little houses on this place. But Granny was now 97 and her health was failing. One day I sensed it was time for her to get ready for heaven. I began to tell her the way there while the house filled with her many children and grandchildren. About three months after she prayed "the sinner's prayer," she exited the world and entered heaven.

There were several killings in the zone, including two of our youthful attenders. A nearby community centre, whose administrator's wife was a Christian, asked us to help their youth keep out of trouble. They provided a free facility which we used for several months. Different church groups visited and put on dramas or special performances with a moral theme. So now we had two areas of ministry and the church continued to grow. We had up to ten motorcycles parked outside as young men came to hear the gospel (and get some dessert!)

We finally realized we would have to get our own church building. But because it was an expensive area, the Assemblies of God had never been able to afford a building. As we looked at the prices, we didn't think we would be able to afford it either. But God had a plan. A former Episcopalian pastor/lawyer had moved to the USA, and rented out his large house to a box-maker. The price was reduced to "only" $75,000. One reason for the low price was an infestation of termites which ruined all the doors, louvered windows, and many of the beams. In a burst of faith I asked the Assemblies of God (AG) in Springfield for a loan of $50,000, and promised another $25,000 from the Anonby Missionary Society. "How will you pay the loan?" asked Dick Nicholson, our AG area director. "We'll get teams, and they will give offerings!" I replied. And that's the way it happened.

As an immediate boost to my step of faith, we got a partial inheritance of $7500 from Charlotte Kinnear, who was a faithful prayer supporter. That was 10% of the price, already paid. Team after team from Canada and Pennsylvania slowly built the church, turning the patio into a sanctuary. The upstairs was a ready-made large parsonage. And we turned the area above the platform into a small suite for ourselves. Now we paid no more rent.

We kept on building, because sometimes you can't stop (especially if you have no walls, no roof and no doors.) We had the promise of three more church groups coming, with construction offerings of about $10,000 each. Meanwhile I got several credit cards offering "0% interest for 12 months." I maxed out those cards to nearly complete the building. Then two groups cancelled and the third was only able to raise $5000. I'd paid off the $50,000 loan, much of it from monthly deductions from my salary.

But now we had an additional US$28,000 construction bill and my credit cards were due. The Dominican Republic government issued new laws that made it very hard to renew our residency. Plus, at age 73, I felt it was time to retire from full time missionary status with the AG. (We became Short Term Missionary Associates.) Ellen had severe back pains with few treatment options in the D.R. I wrote my USA supporters about this construction debt, but there was little response.

What to do? Our AG leaders said "Well, you have a severance/retirement package. You can use that!" Ahem, to get that in bulk meant a huge tax hit, and it still wouldn't cover even half of that debt. How could I pay out $28,000 on my old age pension?

We did some very earnest praying. The Lord gave me an anointing as for three days I did nothing else but pray, phone and email our supporters. And the Lord supplied ALL OF THE $28,000! Miracles all the way! Thank the Lord and thank His faithful servants. The Deep River Church is paid for, and on a recent visit we helped in a campaign with over 500 in a nearby stadium. The church is full, and looking to expand. And all God's blessings on us during our lifetime of ministry happened because with your help we are all "workers together...with God!"

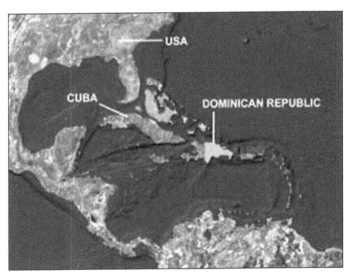

Dominican Republic

New church in Santa Domingo

Typical Dominican national dress

information can be obtained
at ICGtesting.com
in the USA
031949071218
695LV00001B/1/P

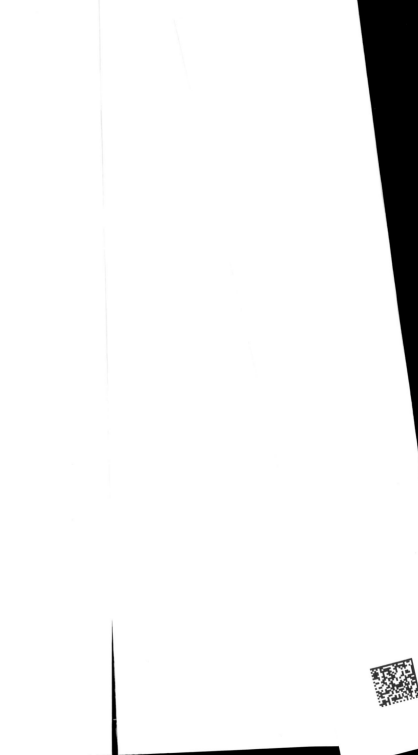